# Practical Handbook on Biodiesel Production and Properties

# Practical Handbook on Biodiesel Production and Properties

Editor

**Rohit Kumar**

**scitus**
academics

**Practical Handbook on Biodiesel Production and Properties**

Edited by **Rohit Kumar**

Printed in 2017

ISBN: 978-1-68117-038-1

Library of Congress Control Number: 2015931839

© 2016 by
SCITUS Academics LLC,
616, Corporate Way, Suite 2, 4766,
Valley Cottage, NY 10989

www.scitusacademics.com

# Contents

vi

# Preface

Biodiesel is a domestically produced, renewable fuel that can be manufactured from vegetable oils, animal fats, or recycled restaurant grease for use in diesel vehicles. Biodiesel's physical properties are similar to those of petroleum diesel, but it is a cleaner-burning alternative. Using biodiesel in place of petroleum diesel, especially in older vehicles, can reduce emissions.

Biodiesel is produced using a broad variety of resources. This diversity has grown significantly in recent years, helping shape a nimble industry that is constantly searching for new technologies and feedstocks. In fact, industry demand for less expensive, reliable sources of fats and oils is stimulating promising research on next-generation feedstocks.

Practical Handbook on Biodiesel Production and Properties focuses on recent trends of biodiesel research, production, and implementation. It also includes practical guidance on the identification of plant resources and their distribution, botanical description, oil extraction, production process, and biodiesel yield. The book will be essential reading for chemists, chemical engineers and agricultural scientists working in both industry and academia on the production of biofuels.

**Editor**

# Chapter 1

# An Investigation of Biodiesel Production from Wastes of Seafood Restaurants

Nour Sh. El-Gendy,[1] A. Hamdy,[1] and Salem S. Abu Amr[2]

[1]Egyptian Petroleum Research Institute, P.O. Box 11727, Nasr City, Cairo, Egypt

[2]Environmental Health Directorate, Ministry of Health, Gaza Strip, Palestine

Received 24 May 2014; Revised 18 September 2014; Accepted 18 September 2014; Published 7 October 2014

## ABSTRACT

This work illustrates a comparative study on the applicability of the basic heterogeneous calcium oxide catalyst prepared from waste mollusks and crabs shells (MS and CS, resp.) in the transesterification of waste cooking oil collected from seafood restaurants with methanol for production of biodiesel. Response surface methodology RSM based on D-optimal deign of experiments was employed to study the significance and interactive effect of methanol to oil M : O molar ratio, catalyst concentration, reaction time, and mixing rate on biodiesel yield. Second-order quadratic model equations were obtained describing the interrelationships between dependent and independent variables to maximize the response variable (biodiesel yield) and the validity of the predicted models were confirmed. The activity of the produced green catalysts was better than that of chemical CaO and immobilized enzyme Novozym 435. Fuel properties of the produced biodiesel were measured and compared with those of Egyptian petro-diesel and international biodiesel standards. The biodiesel produced using MS-CaO recorded higher quality than that produced using CS-CaO. The overall biodiesel characteristics were acceptable, encouraging application of CaO prepared from waste MS and CS for production of biodiesel as an efficient, environmentally friendly, sustainable, and low cost heterogeneous catalyst.

# INTRODUCTION

Development of transesterification process for production of biodiesel as an alternative, green, and sustainable fuel has become an important issue due to diminishing fossil fuel reserves, rising crude oil price, and the stringent exhaust emission regulations [1].

Biodiesel produced by transesterification reaction can be catalyzed with alkali, acid, or enzyme, in which a primary alcohol reacts with the triglycerides of fatty acids form glycerol and esters. Triglyceride for biodiesel production comes from various sources edible and nonedible oil, waste and used oil, and fats [2]. The major hurdle of applicability of biodiesel is the operational cost of its production process, approximately 70–95% of the production cost arose from the feedstock. Homogenous chemical catalyst processes, including alkali or acid, are more practical compared with the enzymatic method. The use of acid catalysts has been found to be used for pretreating high free fatty acid feedstock but the reaction rates for converting triglycerides to methyl esters are very slow. Enzymes have shown good tolerance for the free fatty acid level of the feedstock but the enzymes are expensive. The use of homogenous basic catalytic process overcome the aforementioned problems but suffers some drawbacks: production of large amount of wastewater from washing process of catalyst residues and neutralization step, difficulty of the product separation and purification, and unreusability of the catalysts [3]. Heterogeneous catalytic process overcomes these drawbacks. Among the heterogeneous solid catalysts, calcium oxide CaO has attracted attention due to the elimination of neutralization step, high activity, being active in mild reaction conditions, long catalyst life time, low solubility in methanol, lack of toxicity, ability to withstand high temperatures, ease of recycling, low cost, and being abundantly available in nature as limestone and its performance for biodiesel production is comparable to several homogenous catalysts [4].

In Egypt, millions of liters of waste cooking oil WCO are discarded each year into sewage systems. Thus, it pollutes water streams, causing a lot of waste management problems and, consequently, adds to the cost of treating effluent. Solid wastes density in Egypt averages about $300 \, kg/m^3$, where 60% is organic wastes [5]. Thus, production of biodiesel from WCO using CaO prepared from organic wastes (e.g., eggshells, mollusks shells, crabs shells, etc.) would offer a triple-fact solution: economic, environmental, and waste management.

The aim of this work is to optimize biodiesel transesterification process using waste cooking oil and calcium oxide catalyst prepared from waste mollusks and crabs shells collected from seafood restaurants in an attempt to reach an effective process for practical, low cost industrial biodiesel production.

# MATERIALS AND METHODS

## Materials

Pure calcium oxide as heterogeneous catalyst and methanol (AR Grade) were purchased from Fluka Chemical Corp., Gillingham, UK. Novozym 435 (Candida antarctica Lipase B) was a gift from Novozyme A/S, Bagsvaerd, Denmark, and was supplied as an immobilized enzyme on macroporous acrylic resin. Commercial Egyptian petro-diesel was obtained from a local fuelling station.

# Collection and Preparation of Waste Frying Oil and Heterogeneous Catalyst

The waste frying oil WFO, mollusks, and crabs shells (MS and CS) were collected from local seafood restaurants and prepared according to El-Gendy et al. [3].

The WCO was characterized by high total acid number, density, and viscosity, recording 3 mg KOH/g oil, 0.9208 g/cm$^3$, and 50 cSt, respectively, and its saponification and iodine value were 197 mg KOH/g oil and 119 mgI$_2$/100 g oil, respectively. The WCO consists of ≈24.36, 36.35, 28.68, and 10.61% palmitic (C16:0), stearic (C18:0), oleic (C18:1), and linoleic (C18:2) acid, respectively.

# Catalyst Characterization

The prepared catalysts were characterized according to El-Gendy et al. [3] using differential scanning calorimetric-thermal gravimetric analysis (DSC-TGA) and were performed by Q600 SDT Simultaneous DSC-TGA (New Castle, DE USA), a high-resolution X-ray diffractometer (XRD, PANalytical X'Pert PRO MPD, Netherland) coupled with Cu k radiation source (Å), Dispersive Raman spectrometer (BRUKER-SENTERRA, Germany) equipped with an integral microscope (Olympos), energy dispersive X-ray analysis (EDX, Oxford X-Max, England) conjugated with transmission electron microscope TEM (JEM 2100, Jeol, Japan), analytical Fourier transforms infrared (FT-IR, Perkin Elmer Spectrum One, USA) instrument, scanning electron microscope (SEM, JEOL-model JSM-53000, Japan). Particle sizer model Beckman Coulter Multisizer-3 (Nyon, Switzerland) was used for determination of particle size distribution. The specific surface area of the prepared biocatalysts was measured by Brunauer-Emmett-Teller BET method using low temperature N$_2$adsorption-desorption (NON A3200e, Quantachrome, USA). The samples were tested for pore volume and pore size distribution using Barrett-Joyner-Halenda BJH method. Temperature programmed desorption using CO$_2$ as a probe molecules (CO$_2$-TPD) was used to study basic properties of the prepared biocatalysts and it was done according to Viriya-empikul et al. [1].

# Transesterification

The transesterification reactions were conducted in a laboratory-scale setup, according to El-Gendy et al. [3], and the biodiesel yield was calculated according to Boro et al. [6]. The activity of the prepared biocatalysts was compared with that of commercially available, most effective heterogeneous basic chemical catalyst CaO and immobilized enzyme Novozym 435.

# Experimental Design and Statistical Analysis

Based on D-optimal design of experiments, twenty runs of experiments have been conducted for three levels of four independent variables: methanol:oil M:O (molar ratio; ), catalyst concentration (wt%; ), reaction time (min; ), and mixing rate (rpm; ), to study their effect on the % yield of the produced biodiesel at constant temperature 60°C. MATLAB 7.0 software (MathWorks, USA) was used for experimental design.

Once the experiments were preformed, the next step was to perform a response surface experiment to produce a prediction model to determine curvature, detect interactions among the design factors (independent variables), and optimize the process, that is, to determine the local optimum independent variables with maximum yield of biodiesel. The model used in this study to estimate the response surface is the quadratic polynomial represented by the following equation:

$$Y = \beta_0 + \sum_{i=1}^{n} \beta_i x_i + \sum_{i=1}^{n-1} \sum_{j=i+1}^{n} \beta_{ij} x_i x_j + \sum_{i=1}^{n} \beta_{ii} x_i^2, \tag{1}$$

Where is the biodiesel yield (wt%), is the number of factors, is the intercept term, , , and are the linear, interactive, and quadratic coefficients, respectively. 's are the levels of the independent variables (factors) under study.

The statistical software Design Expert 6.0.7 (Stat-Ease Inc., Minneapolis, USA) was used for regression and graphical analyses of the data obtained and statistical analysis of the model to evaluate the analysis of variance (ANOVA).

## Physicochemical Characterization of the Produced Biodiesel

The produced biodiesel was tested for estimating and evaluating its fuel properties, using the standard methods of analysis for petroleum products, American Society for Testing and Materials ASTM standards methods [7]. The results were compared with the Egyptian standards for petro-diesel and European and American standards of biodiesel (EN14214 and D-6751, resp.) [8, 9].

All the properties were analyzed in two replicates and the final results given below were obtained as the average values.

# RESULTS AND DISCUSSION

## Catalyst Characterization

The thermal transition during the calcination process of each collected waste shells was investigated with TGA/DSC. Figure 1 shows the thermal analysis results along with the weight loss when the temperature was raised from room temperature to 1100°C.

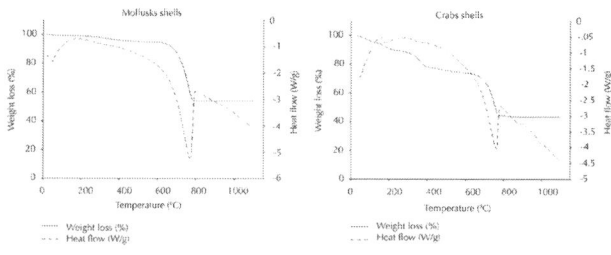

**Figure 1**: TGA/DSC curves.

The thermos gravimetric analysis of MS shows that there was no significant decomposition occurred upon heating up to 600°C, recording weight loss of ≈5%. The dominant decomposition ≈41% weight loss occurred within 700–800°C (peaked at 778) and remained sustained thereafter up to 1100°C. The DSC curve supports the TGA curve, where the heat flow chart illustrates endothermic peak at 778°C. This reveals the production of new compound.

Upon calcination of CS, the TGA analysis reveals three phases of mass loss with a total weight loss of ≈56.77%. The first mass loss was observed at temperatures below 200°C, recording weight loss of ≈5% (peaked at 170°C). The second mass loss occurred within 250–400°C, recording weight loss of ≈20% (peaked at 320°C). The dominant decomposition with ≈30% weight loss occurred within 650–770°C (peaked at 750°C) and remained sustained thereafter up to 1100°C. The DSC curve shows an exothermic peak at 130°C. According to Boro et al. [6], this can be attributed to evaporation of water, crystallization, decomposition of organic components, or possible rearrangement in the structural arrangement within the compound itself. The DSC curve shows also a weak broad endothermic peak located at the range 320–370°C and another sharp endothermic peak at 753°C. The two endothermic peaks might indicate the decomposition of compounds and formation of a new one.

According to N. B. Singh and N. P. Singh [10], the weight loss might be attributed to the decomposition of $CaCO_3$ through the loss of carbon dioxide $CO_2$ and production of calcium oxide CaO and also due to the possible removal of absorbed water molecules which occurs according to the following dissociation equation:

$$CaCO_3(S) \rightarrow CaO(S) + CO_2(g) \qquad (2)$$

It was observed that during calcination, the MS and CS turned completely pale grey and white, respectively. According to Engin et al. [11], this indicates that calcium carbonate $CaCO_3$ is converted to CaO with elevating calcination temperatures.

The XRD patterns of natural MS (Figure 2) are mainly aragonite $CaCO_3$ (JCPDS card number: 024-0025) and upon calcination at 400°C, it was changed to calcite $CaCO_3$ (JCPDS card number: 005-

0586) and remained unchanged up to calcination at 600°C, while calcination at 700°C showed mixture of calcite $CaCO_3$ (JCPDS card number: 005-0586) and lime CaO (JCPDS card number: 043-1001). The XRD patterns showed a pure crystalline CaO (JCPDS card number: 043-1001) for calcination temperatures 800–1100°C, while the XRD patterns of CS revealed that the natural CS is composed mainly of crystalline calcite $CaCO_3$ (JCPDS card number: 005-0586) and also to a minor extent of monohydrocalcite $CaCO_3H_2O$ (JCPDS card number: 022-0147). Upon calcination at 400°C, $CaCO_3H_2O$ was lost and only calcite $CaCO_3$ (JCPDS card number: 005-0586) detected and remained unchanged up to calcination at 600°C. The XRD patterns showed a pure crystalline calcia or burnt lime CaO (JCPDS card number: 004-0777) for calcination temperatures 700–1100°C. The obtained XRD patterns could explain the TGA/DSC flowcharts of CS, where the weak broad endothermic peak at the range of 320–370°C might be due to the loss of water and decomposition of $CaCO_3H_2O$ and the sharp one at 753°C might be due to the decomposition of $CaCO_3$ and formation of CaO.

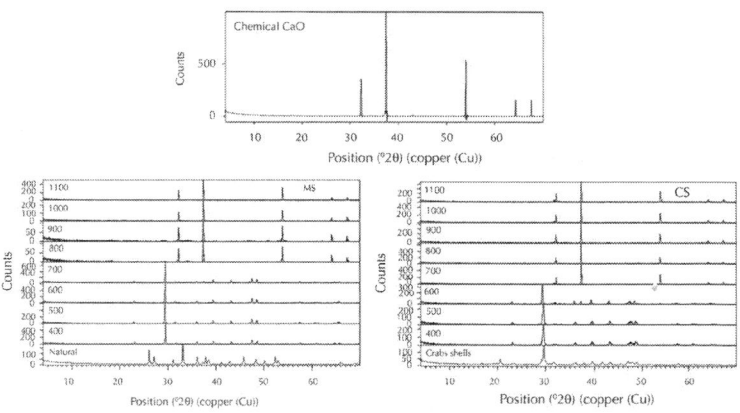

**Figure 2**: XRD patterns during calcination at different temperatures.

Narrow and high intense peaks of the calcinated MS and CS (at temperature ≥800 and 700°C, resp.) could define the well-crystallized nature of the prepared biocatalyst similar observation which was reported by Boey et al. [2] and Viriya-empikul et al. [12] on preparation of biocatalyst from waste mud crab shells (Scylla serrata) and waste mollusk shells for biodiesel production from palm olein oil.

The crystal size was calculated from the XRD data (Table 1), recording ≈42.15 nm for natural MS, which decreased upon calcination at 800°C, recording ≈38.77 nm. Similar observation was reported by Yoosuk et al. [13] who attributed this to the presence of water molecules during calcination of $CaCO_3$ to CaO. But, the crystalline size of the uncalcinated CS was found to be ≈32.15 nm and it was slightly increased by calcination at 700°C, recording ≈32.82 nm. Borgwardt [14] and Liu et al. [15] reported that calcination of $CaCO_3$ leads to formation and growth of CaO. When $CaCO_3$ decomposes at high temperatures, small CaO grains formed and then the contact grains form necks and begin to grow resulting in an increase in the average grain size.

**Table 1**: Some characteristics of the prepared biocatalysts

| Parameters | Natural MS | MS at 800°C | Natural CS | CS at 700°C |
|---|---|---|---|---|
| Crystal size nm | 42.15 | 38.77 | 32.15 | 32.82 |
| Average particle size m | 11.07 | 9.86 | 10.27 | 8.24 |
| $S_{bet}$ m$^2$/g | — | 12.63 | — | 43.73 |
| Pore volume cm$^3$/g | — | 0.046 | — | 0.211 |
| Pore size nm | — | 0.94 | — | 1.84 |
| Basicity mmol $CO_2$/g | — | | — | 27 |

Raman spectra (Figure 3) are consistent with the XRD patterns. In case of natural MS, the vibration bands at 155, 197, 650, 703, and 1086 cm$^{-1}$ correspond to those of aragonite $CaCO_3$. The vibration bands at 360, 1075, 1334, and 1463 cm$^{-1}$ of calcined MS at 800°C correspond to pure lime CaO, while the vibration bands at 155, 282, 714, and 1087 cm$^{-1}$ of natural CS correspond to calcite $CaCO_3$. The high vibration bands at 360, 960, and 1075 cm$^{-1}$ of calcined CS at 700°C correspond to pure burnt lime CaO.

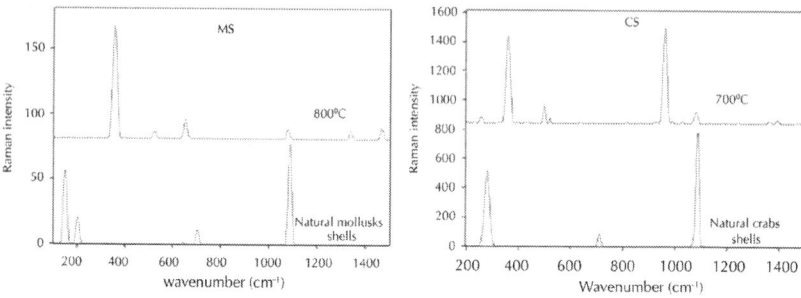

**Figure 3**: Raman spectra at different calcination temperatures.

EDX analysis confirmed also the results obtained from XRD and Raman. The EDX analysis (Table 2) revealed that the chemical composition of the shells was highly affected by calcination. The uncalcinated shells exhibit oxygen as the main component, 56.62 and 53.69%, wt%, while in calcinated shells at 800°C and 700°C, calcium represents the major component (62.32 and 49.89%, wt%) for MS and CS, respectively. Similar observation was reported by Viriya-empikul et al. [12] and Birla et al. [16], where the main component in the calcinated shells was calcium and other elements, for example, Na, Mg, and so forth, were found in trace amounts. Stoichiometrically, this is true, as in CaO, Ca is the main constituent ($\approx$71%), while oxygen is the major one in $CaCO_3$ ($\approx$48%).

**Table 2**: Chemical composition of biomass derived catalysts

| Biomass | Chemical composition (wt%) |
|---|---|
| Natural mollusks shells | C (17.35%) Mg (0.01%) Ca (26.01%) O (56.62%) |
| Calcinated mollusks shells at 800°C | C (3.47%) Mg (0.05%) Ca (62.32%) O (34.16%) |
| Natural crabs shells | C (14.44%) Na (1.1%) Mg (0.92%) Ca (29.85%) O (53.69%) |
| Calcinated crabs shells at 700°C | C (7.92%) Na (0.03%) Mg (0.53%) Ca (49.89%) O (41.63%) |

The basicity of the prepared CaO from MS and CS recorded 52.18 and 27 mmol $CO_2$/g, respectively. This might affect the activity of the catalyst in the transesterification reaction, where the mechanism of CaO in transesterification reactions would be started by dissociation of CaO (Scheme 1). Then oxide anion attacks the methanol to form methoxide anion, which is why excess methanol is required to drive the reaction in the forward direction. The methoxide anion attacks the carbonyl carbon of the triglyceride to form a tetrahedral intermediate. The rearrangement of the intermediate molecule forms a mole of methyl ester and diglyceride. Another methoxide anion attacks the carbonyl carbon in the formed diglyceride, forming another mole of methyl ester and monoglyceride. Finally, a new methoxide anion attacks the monoglyceride, producing a total of three moles of methyl esters and a mole of glycerol.

$$CaO \quad Ca^{2+} \quad O^{2-}$$
$$O^{2-} + CH_3OH \quad OH^- + CH_3O^-$$
$$OH^- + CH_3OH \quad H_2O + CH_3O^-$$

**Scheme 1:** The FTIR patterns of MS and CS with respect to calcination process (Figure 4) were nearly the same, showing major peaks around 1420–1471, 860–874, and 709 cm$^{-1}$ in patterns of natural shells, which disappeared in that of calcinated shells. According to Engin et al. [11], these peaks are attributed to asymmetric stretch; out-of-plane bend and in-plane bend vibration modes for molecules. The strong sharp peak at 3643 cm$^{-1}$ in uncalcinated CS confirms the XRD pattern and presence of $CaCO_3H_2O$. Also, upon calcination, weak bands around 2509 and 1786–1793 cm$^{-1}$ disappeared and new weak peak appeared at 1111 cm$^{-1}$ for calcinated MS and 1089 and 1056 cm$^{-1}$ for calcinated CS. The observed changes in IR patterns might indicate the complete transformation of $CaCO_3$ to CaO. Similar observation was reported by Roschat et al. [17].

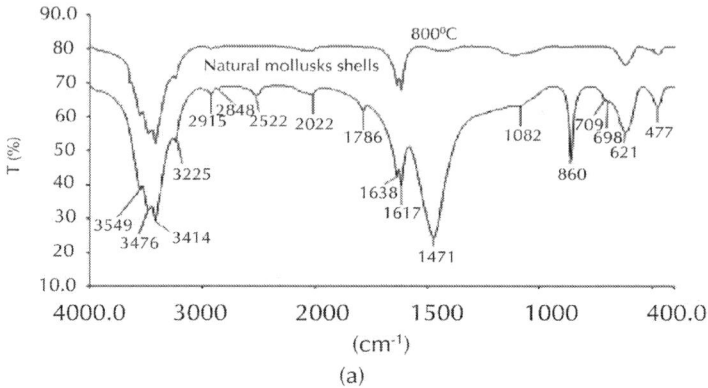

(a)

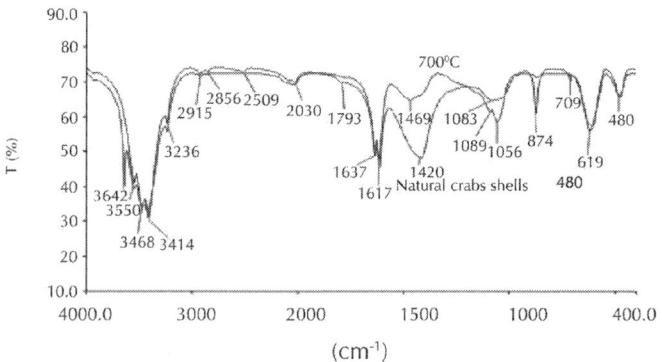

(cm⁻¹)

**Figure 4**: FTIR spectra of mollusks shells (a) and crabs shells (b) at different calcination temperatures.

In the viewpoint of preparation time, energy consumption, and cost of catalyst preparation, the temperatures of 800°C and 700°C were selected as perfect calcination temperature to prepare biocatalysts from natural waste mollusks and crabs shells, respectively.

The morphology of natural and calcined shells was investigated by SEM at equal magnification of 500x. SEM micrograph of catalyst derived from MS (Figure 5(b)) shows that, upon calcination, the morphology of MS changed from layered bulky substances without any clear pores on its surface (Figure 5(a)) to porous particles of various sizes and shapes, with higher specific surface area. The SEM micrographs of the natural crabs shells (Figure 5(c)) showed bulky and nonuniform clustered substances with clear pores on its surface which transformed to relatively similar aggregates of porous smaller particles with higher specific surface area upon calcination at 700°C (Figure 5(d)). This porosity is probably due to the fact that a large number of gaseous water molecules are released upon the decomposition of $CaCO_3H_2O$. Hu et al. [18] reported that the gaseous water molecules create high porosity in the catalyst, that is, to act as porogens.

(a)

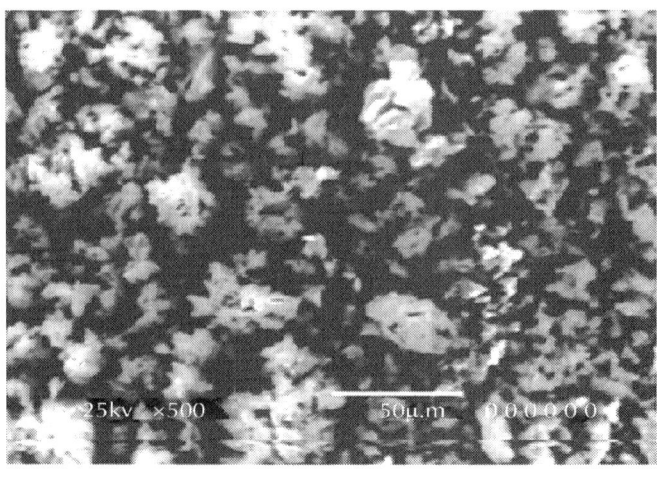

(b)

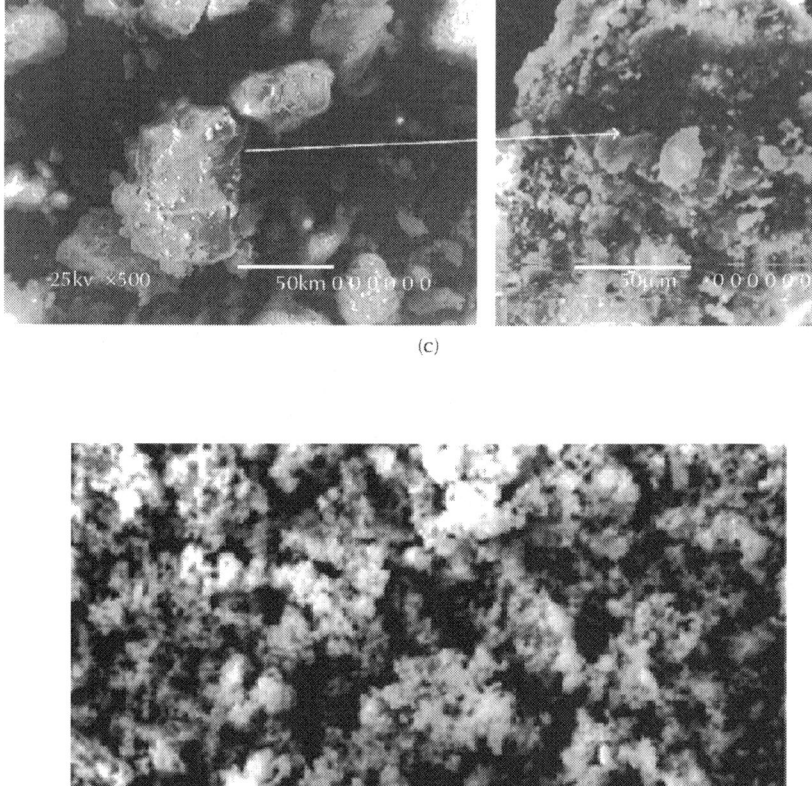

(c)

(d)

**Figure 5**: SEM images of natural mollusks shells (a), calcined mollusks shells at 800°C (b), natural crabs shells, (c) and calcined crabs shells at 700°C (d).

The particles size distribution (Table 1) proved the SEM analysis (Figure 5), where a large part of the particles size distribution of the natural MS and CS was within the size range of 8.58–17.23 and

9.01–18.43 µm, while the rest was within the range of 6.48–7.13 and 6.52–7.25 µm, with overall average particles diameter of ≈11.07 and 10.27 µm, respectively. But upon calcination at 800 and 700°C the particles size decreased, where a large part of the particles size distribution was in the range of 7.93–14.50 and 7.71–10.74 µm and the rest was within the range of 6.37–6.80 and 6.38–6.81 µm, with overall average particles diameter of ≈9.86 and 8.24 µm, respectively. The smaller size of the grains and aggregates would provide higher specific surface areas. Because the prepared biocatalyst has relatively large particle sizes, it is easy to separate the catalyst from the products after the reaction, by filtration or centrifugation.

The BET surface area $S_{BET}$ of the prepared catalysts (Table 1) recorded 12.63 and 43.73 m²/g for CaO prepared from MS and CS, respectively. This coincides with the SEM observation (Figures 5(b) and 5(d)). The BJH method was used for calculations of the pore size distributions (Table 1) and the prepared biocatalyst recorded total pore volume of ≈0.046 and 0.211 cm³/g, respectively, and the pore size, distributed between ≈0.83 and 17.18 nm and between 1.04 and 15.58 nm, with average pore diameter of 0.94 and 1.84 nm, respectively, indicating that most of the pore size distributions are found in the microporous range. According to Roschat et al. [17], a high porosity catalyst is a key requirement to achieve high conversion efficiency for heterogeneous process, thereby high surface area or high catalytic sites are necessary. Sharma et al. [19] also reported a high pore size to be desirable for better diffusion of reactants and product molecules. Thus, this would recommend the biocatalyst prepared from MS and CS for application in biodiesel production.

The $N_2$ adsorption-desorption isotherms of the prepared CaO from MS and CS are shown in Figures 6(a) and 6(b) and they are of type II isotherm (based on IUPAC's classification) with a low slope in the middle region of the isotherm and a desorption curve almost overlapping with adsorption curve. Applying BET equation, and 0.9482 (<1) and and 17.012 (>1), respectively, indicating monolayer formation. The positive curvature at the lowest pressures indicates a distribution of adsorption energies. The isotherms have a hysteresis loop of H3 and slopping adsorption and desorption branches covering a large range of with underlying type II isotherm.

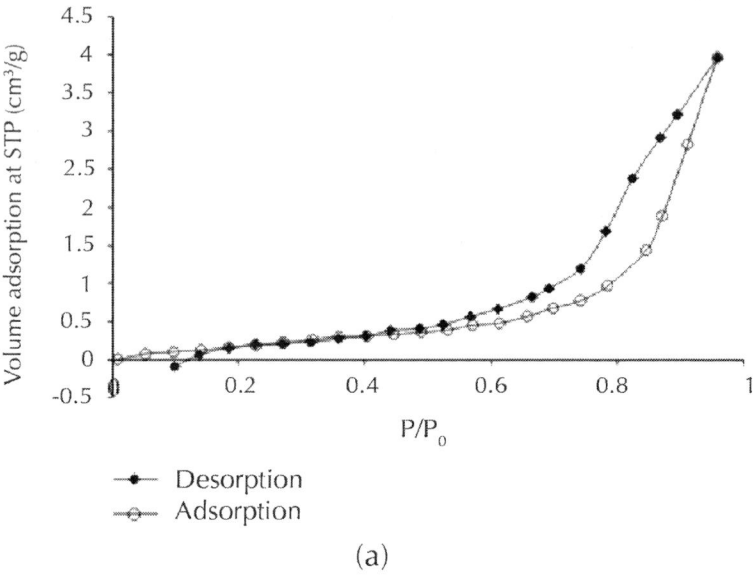

(a)

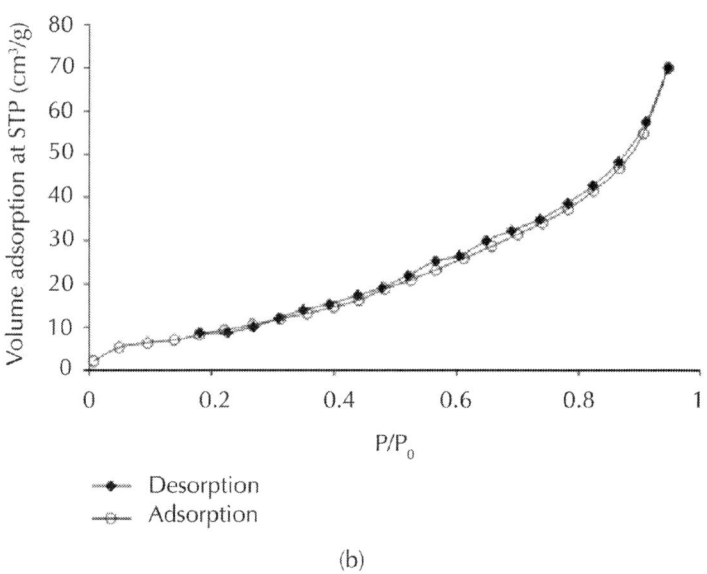

(b)

**Figure 6**: $N_2$ adsorption-desorption isotherm of calcined MS at 800°C (a) and CS at 700°C.

## Regression Model and Its Validation

The complete design matrix with experimental and predicted values of % yield of biodiesel using different CaO catalyst prepared from mollusks and crabs shells is presented in Table 3. Based on D-optimal design and experimental data, two second-order quadratic models for the applied catalysts have been predicted and can be given as follows:

$$Y_1 = 82.43 - 2.21A - 0.98B + 3.12C + 0.80D + 4.48A^2$$
$$+1.89B^2 - 7.51C^2 - 2.56D^2 - 2.66AB - 1.16AC \tag{3}$$
$$+5.79AD + 2.50BC + 8.42BD + 0.87CD,$$
$$Y_2 = 85.05 + 3.41A - 3.28B - 5.05C - 0.6D$$

$$-2.19A^2 + 4.63B^2 + 1.90C^2 + 0.04D^2 + 4.89AB \tag{4}$$
$$-0.56AC + 0.32AD - 3.194BC - 5.45BD + 2.53CD,$$

where $Y_1$ and $Y_2$ are the biodiesel yields from transesterification processes using biocatalyst prepared from MS and CS, respectively.

**Table 3:** Experimental design matrix with experimental and predicted values of biodiesel yield

| Run number | Factors/levels | | | | | | | | | Biodiesel yield (wt%) | | | |
| --- | --- | --- | --- | --- | --- | --- | --- | --- | --- | --- | --- | --- | --- |
| | M:O (Molar ratio) | | Catalyst (wt%) | | Time (min) | | Mixing rate (rpm) | | | CaO from mollusks shells | | CaO from crabs shells | |
| | Levels | Actual value | Levels | Actual value | Levels | Actual value | Levels | Actual value | | Experimental | Predicted | Experimental | Predicted |
| 1 | 3 | 12:1 | 3 | 9 | 3 | 120 | 2 | 300 | | 79.80 | 79.90 | 85 | 85.60 |
| 2 | 1 | 6:1 | 3 | 9 | 3 | 120 | 1 | 200 | | 85.10 | 85.10 | 74 | 74.00 |
| 3 | 2 | 9:1 | 3 | 9 | 2 | 60 | 3 | 400 | | 87.00 | 87.00 | 82.5 | 82.50 |
| 4 | 1 | 6:1 | 3 | 9 | 1 | 30 | 2 | 300 | | 78.40 | 78.40 | 85.5 | 85.50 |
| 5 | 1 | 6:1 | 2 | 6 | 3 | 120 | 3 | 400 | | 79.20 | 79.20 | 78.5 | 78.50 |
| 6 | 2 | 9:1 | 3 | 9 | 1 | 30 | 3 | 400 | | 76.00 | 76.00 | 88 | 88.00 |
| 7 | 2 | 9:1 | 2 | 6 | 1 | 30 | 2 | 300 | | 71.80 | 71.80 | 92 | 92.00 |
| 8 | 3 | 12:1 | 1 | 3 | 1 | 30 | 1 | 200 | | 83.40 | 83.40 | 91 | 90.85 |
| 9 | 1 | 6:1 | 1 | 3 | 1 | 30 | 1 | 200 | | 91.50 | 91.75 | 93 | 93.50 |
| 10 | 1 | 6:1 | 1 | 3 | 2 | 60 | 2 | 300 | | 88.00 | 87.90 | 92.5 | 92.90 |
| 11 | 2 | 9:1 | 2 | 6 | 3 | 120 | 1 | 200 | | 73.80 | 73.80 | 80 | 80.00 |
| 12 | 2 | 9:1 | 1 | 3 | 3 | 120 | 2 | 300 | | 78.40 | 78.40 | 93 | 93.00 |
| 13 | 3 | 12:1 | 1 | 3 | 3 | 120 | 3 | 400 | | 78.50 | 78.65 | 96 | 96.50 |
| 14 | 3 | 12:1 | 2 | 6 | 1 | 30 | 3 | 400 | | 78.40 | 78.40 | 91 | 91.00 |
| 15 | 3 | 12:1 | 2 | 6 | 2 | 60 | 1 | 200 | | 74.20 | 74.35 | 89 | 89.50 |
| 16 | 3 | 12:1 | 3 | 9 | 3 | 120 | 2 | 300 | | 80.00 | 79.90 | 86.2 | 85.60 |
| 17 | 1 | 6:1 | 1 | 3 | 1 | 30 | 1 | 200 | | 92.00 | 91.75 | 94 | 93.50 |
| 18 | 3 | 12:1 | 2 | 6 | 2 | 60 | 1 | 200 | | 74.50 | 74.35 | 90 | 89.50 |
| 19 | 3 | 12:1 | 1 | 3 | 3 | 120 | 3 | 400 | | 78.80 | 78.65 | 97 | 96.50 |
| 20 | 1 | 6:1 | 1 | 3 | 2 | 60 | 2 | 300 | | 87.80 | 87.90 | 93.3 | 92.90 |

Positive sign in front of the terms indicate synergetic effect, whereas negative sign indicates antagonistic effect.

Pareto charts, which are very useful in design of experiments, were used in this work to make it much easier to visualize the main and interaction effects of all factors to the response variable that is biodiesel yield (Figure 7).

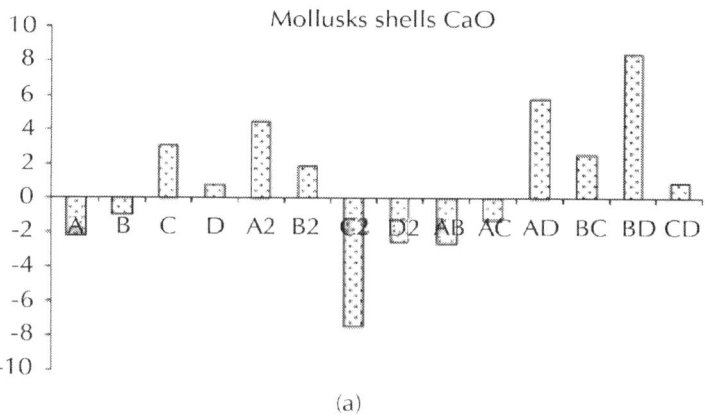

(a)

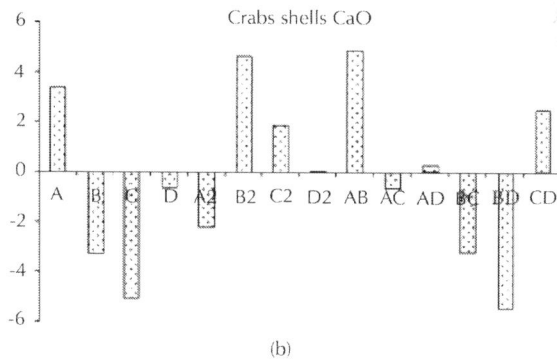

(b)

**Figure 7**: Pareto chart showing the effects of different independent variables on % biodiesel yield.

In case of using biocatalyst prepared from MS in the transesterification process, reaction time has a high positive impact on the reaction yield, followed by the mixing rate. But, M : O molar ratio has a high negative impact followed by catalyst loading. The interaction effects of M:O molar ratio and catalyst loading have a higher negative impact than M:O and reaction time on the biodiesel yield, while the interactive effect of catalyst loading and mixing rate expresses a very high positive impact on the biodiesel yield, coming after it the interaction between M:O ratio and mixing rate, catalyst loading, and reaction time, in a decreasing order, and the interactive effect of reaction time and mixing rate has a low positive impact on the reaction yield.

In case of using biocatalyst prepared from CS in the transesterification process, only the M:O molar ratio has positive impact; that is, increasing the M:O would increase the biodiesel yield. But, reaction time has a high negative impact followed by catalyst concentration and mixing rate, in a decreasing order; that is, their increase would lower the biodiesel yield. The positive interactive impact on the biodiesel yield was expressed by M:O and catalyst loading, reaction time, and mixing rate and M:O and mixing rate with a decreasing order. While the interaction between catalyst loading and mixing rate, catalyst loading, and reaction time and M:O and reaction time has negative impact on the biodiesel yield in a decreasing order.

The validity of the fitted models (3) and (4) was evaluated and their statistical significance was controlled by -test. The analyses of variance (ANOVA) for the response surface full quadratic models are given in Tables 4 and5. It can be indicated that the models are highly statistically significant at 95% confidence level, with value of 940.12 and 99.58 with very low probability value of <0.0001, respectively; that is, there is less than 0.01% chance that this error is caused by noise. The values of the determination coefficients, and , which measure the model fitting reliability for the models (3) and (4) were calculated to be (0.9996, 0.9986) and (0.9964 and 0.9864), respectively. This suggests that approximately 99.96% and 99.64% of the variance is attributed to the variables and indicated a high significance of the predicted models. Thus, only 0.04% and 0.36% of the total variations cannot be explained by the models, respectively, which ensures the good adjustment of the above models to the experimental data. Confirmation of the adequacy of the regression models was reflected also by the good agreement between experimental and predicted values of response variables as shown in Table 3, where the actual biodiesel yield for mollusks and crabs shells biocatalyst ranged from 71.80 to 92% and from 74 to 97%

and there corresponding predicted values are 71.80 and 91.75% and 74 and 96.50%, respectively. "Adeq Precision" measures the signal-to-noise ratio. A ratio greater than 4 is desirable. The ratio of 102.01 and 36.452, respectively, indicated an adequate signal. These models are reliable and can be used to navigate the design space.

**Table 4**: Analysis of variance of the fitted quadratic regression model (3)

| Source | SS* | df* | MS* | value | value | Remarks |
|--------|-----|-----|-----|-------|-------|---------|
| Model | 671.25 | 14 | 47.95 | 940.12 | <0.0001 | Highly significant |
| A | 46.97 | 1 | 46.97 | 920.97 | <0.0001 | Highly significant |
| B | 3.06 | 1 | 3.06 | 59.98 | 0.0006 | Highly significant |
| C | 18.50 | 1 | 18.50 | 362.76 | <0.0001 | Highly significant |
| D | 2.14 | 1 | 2.14 | 41.94 | 0.0013 | Significant |
| $A^2$ | 16.46 | 1 | 16.46 | 322.79 | <0.0001 | Highly significant |
| $B^2$ | 2.63 | 1 | 2.63 | 51.52 | 0.0008 | Highly significant |
| $C^2$ | 52.85 | 1 | 52.85 | 1036.26 | <0.0001 | Highly significant |
| $D^2$ | 5.07 | 1 | 5.07 | 99.36 | 0.0002 | Highly significant |
| AB | 6.80 | 1 | 6.80 | 133.27 | <0.0001 | Highly significant |
| AC | 1.84 | 1 | 1.84 | 36.17 | 0.0018 | Significant |
| AD | 33.02 | 1 | 33.02 | 647.50 | <0.0001 | Highly significant |
| BC | 8.84 | 1 | 8.84 | 173.38 | <0.0001 | Highly significant |
| BD | 28.39 | 1 | 28.39 | 556.71 | <0.0001 | Highly significant |
| CD | 1.08 | 1 | 1.08 | 21.19 | 0.0058 | Significant |
| Residual | 0.26 | 5 | 0.051 | | | |
| Corrected total | 671.50 | 19 | | | | |

**Table 5**: Analysis of variance of the fitted quadratic regression model (4)

| Source | SS* | df* | MS* | value | value | Remarks |
|--------|-----|-----|-----|-------|-------|---------|
| Model | 708.18 | 14 | 50.58 | 99.58 | <0.0001 | Highly significant |
| A | 111.75 | 1 | 111.75 | 219.98 | <0.0001 | Highly significant |
| B | 34.40 | 1 | 34.40 | 67.72 | 0.0004 | Highly significant |
| C | 48.63 | 1 | 48.63 | 95.73 | 0.0002 | Highly significant |
| D | 1.20 | 1 | 1.20 | 2.37 | 0.1845 | Nonsignificant |
| $A^2$ | 3.92 | 1 | 3.92 | 7.73 | 0.0389 | Possibly significant |
| $B^2$ | 15.77 | 1 | 15.77 | 31.05 | 0.0026 | Significant |
| $C^2$ | 3.39 | 1 | 3.39 | 6.66 | 0.0493 | Possibly significant |

| $D^2$ | 9.279E-004 | 1 | 9.279E-004 | 1.827E-003 | 0.9676 | Nonsignificant |
|---|---|---|---|---|---|---|
| AB | 22.96 | 1 | 22.96 | 45.20 | 0.0011 | Significant |
| AC | 0.43 | 1 | 0.43 | 0.84 | 0.4021 | Nonsignificant |
| AD | 0.10 | 1 | 0.10 | 0.20 | 0.6720 | Nonsignificant |
| BC | 14.40 | 1 | 14.40 | 28.35 | 0.0031 | Significant |
| BD | 11.88 | 1 | 11.88 | 23.39 | 0.0047 | Significant |
| CD | 9.21 | 1 | 9.21 | 18.13 | 0.0080 | Significant |
| Residual | 2.54 | 5 | 0.51 | | | |
| Corrected total | 710.72 | 19 | | | | |

The relationship between predicted and experimental values of biodiesel yield for CaO prepared from MS and CS is shown in Figures 8(a) and 8(b). It can be seen that there is a high correlation ($R^2 \gg 1$) between the predicted and experimental values indicating that the predicted and experimental values were in high reasonable agreement. It means that the data fit well with the model and give a convincingly good estimate of response for the system in the experimental range studied.

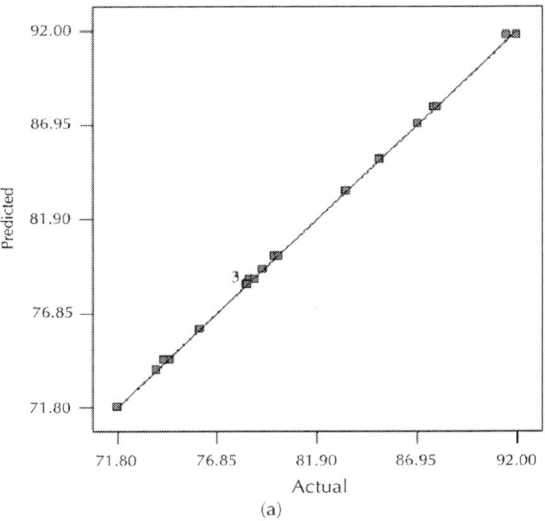

(a)

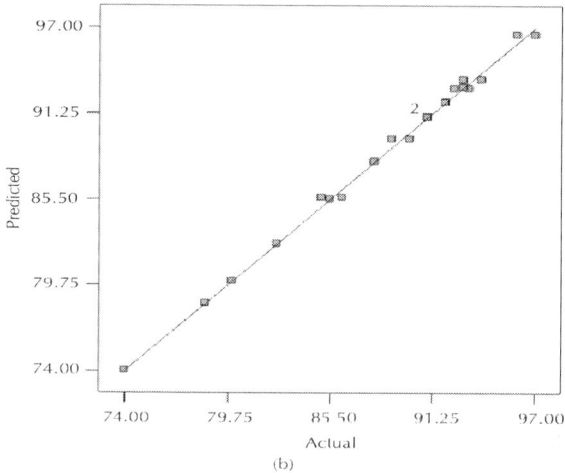

(b)

**Figure 8**: Actual versus predicted biodiesel yield using mollusks (a) and crabs (b) shells-CaO.

The perturbation plots in Figures 9(a) and 9(b) show the comparative effects of all independent variables on the biodiesel yield. The curvatures of the four factors from the center point confirm the statistical data obtained from analysis of variance (ANOVA, Tables 4 and 5), that is, the significance of each parameter (coefficient). In case of CaO prepared from MS (Figure 9(a)), the sharp curvature of the two factors, M:O () and reaction time (), shows that the response biodiesel yield was very sensitive to these two variables. The comparatively low curvature of catalyst concentration () and mixing rate () curves shows less sensitivity of biodiesel yield towards the change in these two factors. The curvatures also confirm the data illustrated in Pareto chart (Figure7(a)), where the increase of M:O molar ratio decreases the biodiesel yield, while the increase in reaction time increases the biodiesel yield. The sensitivity of biodiesel yield towards the four variables can be ranked in the following decreasing order reaction time > M:O > catalyst concentration > mixing rate. In case of CaO prepared from CS (Figure 9(b)), all the parameters, within the studied range, have highly significant effect on the biodiesel yield except the mixing rate () which has relatively nonsignificant effect. The curvatures also confirm the data illustrated in Pareto chart (Figure 7(b)), where the increase of M:O molar ratio increases the biodiesel yield, while the increase of reaction time, catalyst loading, and mixing rate decrease the yield, with decrease of sensitivity (reaction time > M:O ≈ catalyst concentration ≫ mixing rate). These observations are well matched to the model mathematical equations (3) and (4).

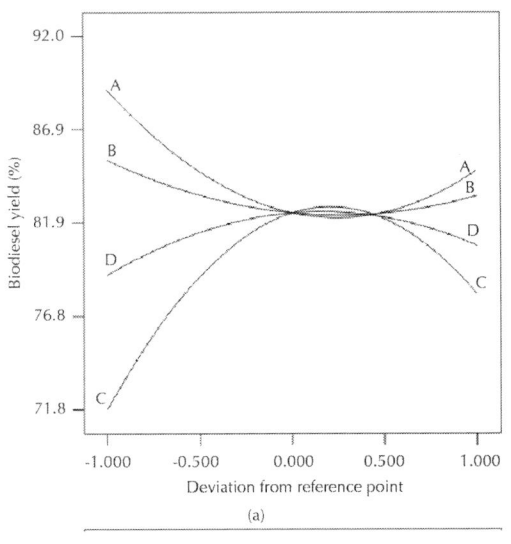

(a)

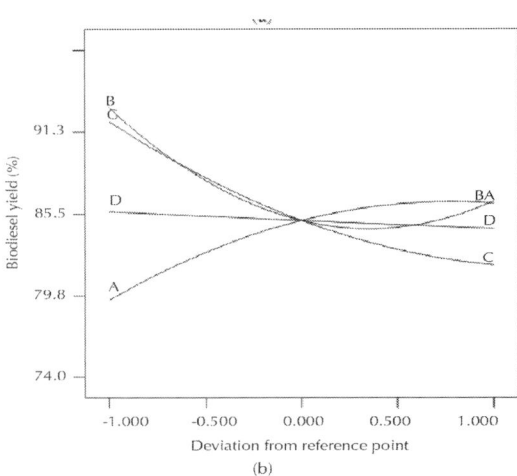

(b)

**Figure 9**: Perturbation plot for biodiesel yield in case of mollusks (a) and crabs (b) shells-CaO.

# Response Surface Methodology

Three-dimensional response surface graphical diagrams of the regression equations (3) and (4) were plotted (Figures 10 and 11) to understand the interactive relationship between the independent variables and % yield of biodiesel and determine the optimum conditions for maximum biodiesel production, using CaO prepared from MS and CS, respectively.

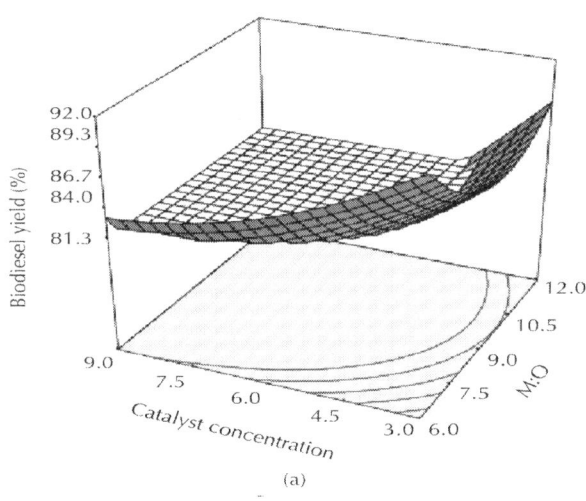

(a)

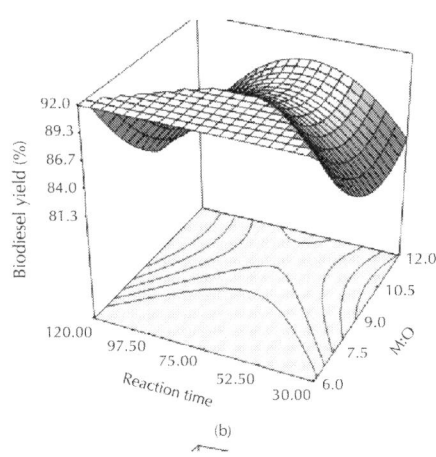

(b)

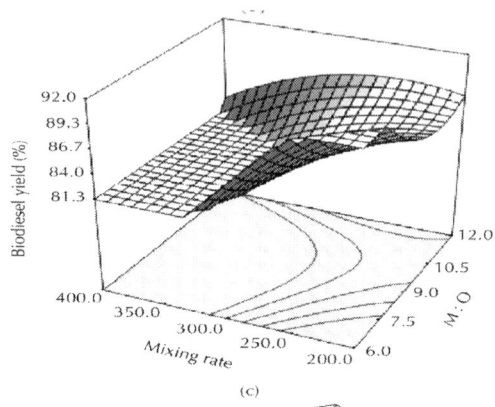

(c)

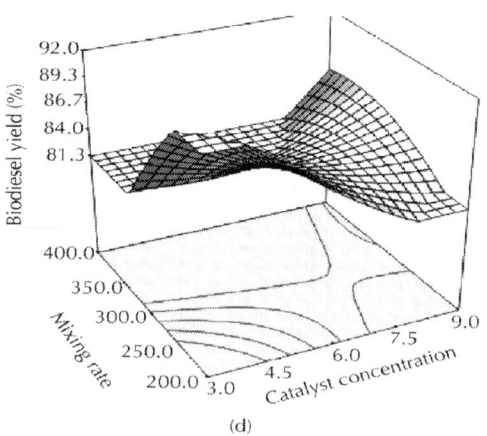

(d)

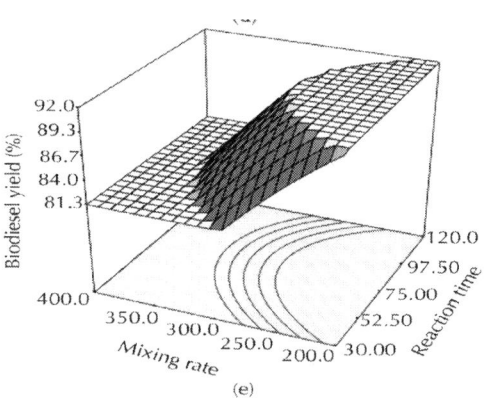

(e)

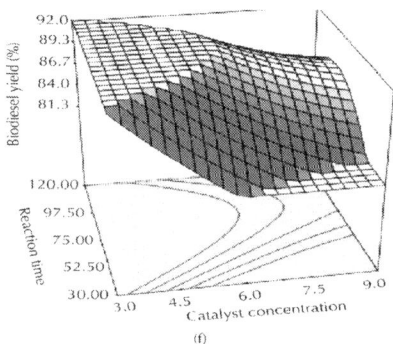

(f)

**Figure 10**: Response surface plots of biodiesel yield using CaO prepared from waste mollusks shells.

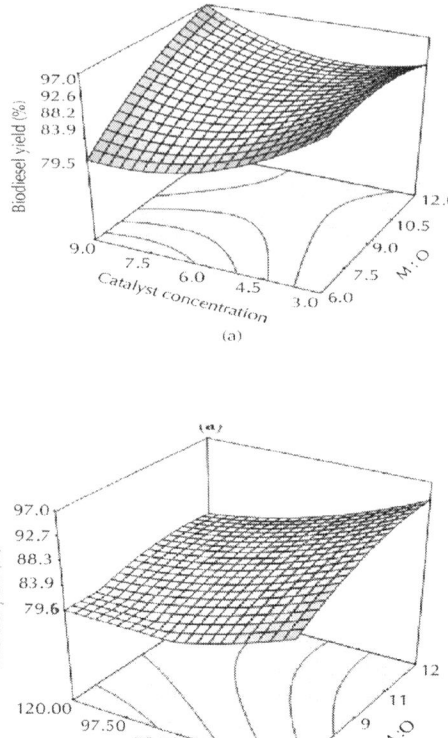

(a)

(b)

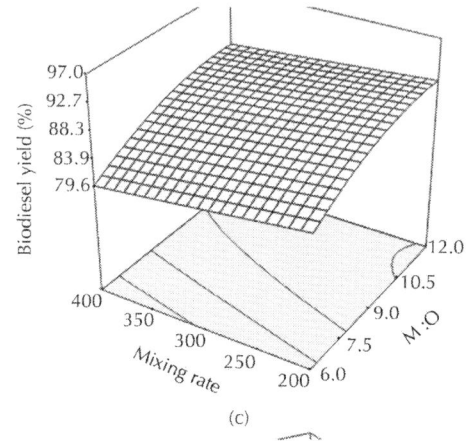

(c)

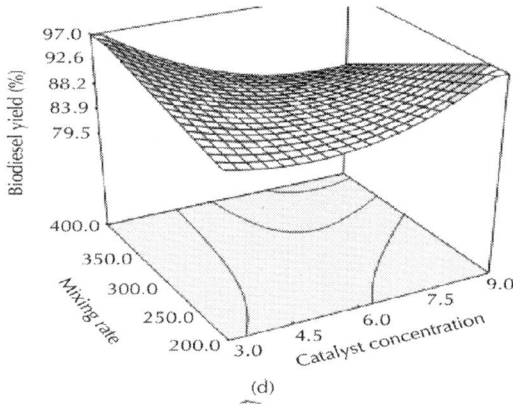

(d)

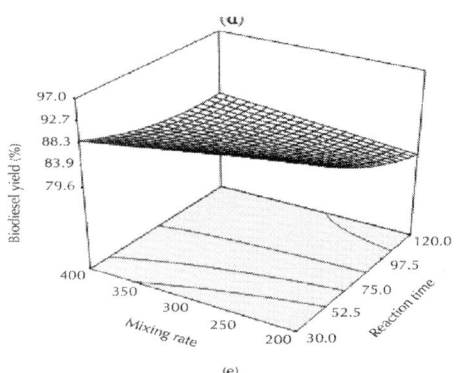

(e)

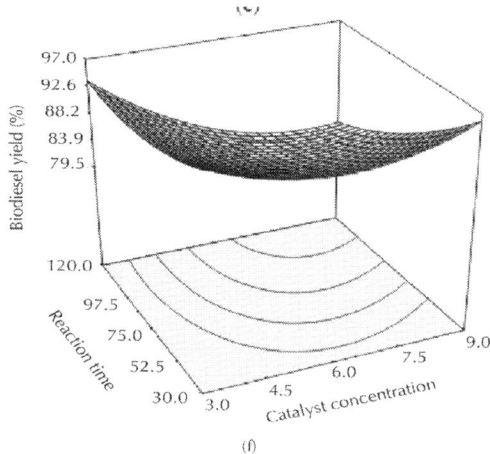

**Figure 11**: Response surface plots of biodiesel yield using CaO prepared from waste crabs shells.

Figure 10(a) illustrates the effect of M : O molar ratio and the prepared MS-CaO concentration on biodiesel production at a constant reaction time of 30 min and mixing rate of 200 rpm. The increase of M : O ratio and catalyst loading decreased the biodiesel yield. The maximum yield ≈92% occurred within the range 6 : 1–6.5 : 1 M : O and 3–4.5 wt% catalyst concentration, but further increase in M : O and catalyst slightly decreased the biodiesel yield to ≈84% at 6 : 1 and 9 wt%, 89% at 12 : 1 and 6 wt%, and 81% at 12 : 1 and 9 wt%. M : O higher than the stoichiometric ratio has generally been adopted for biodiesel production. But further increase in M : O would lower the biodiesel yield as it might have dilution effect on the catalyst concentration. Glycerol would dissolve in excessive methanol and subsequently inhibit the reaction of methanol with oil and catalyst. Also the separation of glycerol would be difficult, which would consequently shift the equilibrium in the reverse direction [20].

Figure 10(b) represents the effect of M : O molar ratio and reaction time on biodiesel production at a constant catalyst concentration 3 wt% and mixing rate 200 rpm. It seems within the studied range of experiments, the increase of reaction time increased the biodiesel yield, recording ≈92% at 6 : 1–6.5 : 1 M : O and 75–93 min molar ratio. The high basicity of MS-CaO enabled transesterification in short reaction time. But the biodiesel yield slightly decreased with further increment of M : O molar ratio and time reaching ≈86% at 12 : 1 M : O and 120 min. The decrease in biodiesel production at higher reaction time might be due to the saturation of the active sites of the catalyst.

Figure 10(c) represents the effect of M : O and mixing rate on biodiesel production at a constant reaction time 30 min and MS-CaO concentration 3 wt%. Biodiesel production recorded its maximum yield ≈92% at 6 : 1–6.5 : 1 M : O and 200–250 rpm, further increment to 12 : 1 M : O and 400 rpm, decreased the yield to ≈86%.

Figure 10(d) shows the interactive effect of mixing rate and CaO concentration. It is obvious from the 3D plot the high positive interaction of these factors, at low catalyst loading and high rpm, the biodiesel yield increased recording ≈81% at 3 wt% and 400 rpm and also at high catalyst loading 9 wt% and low mixing rate 200 rpm. But at low mixing rate 200–250 rpm and catalyst loading 3 wt%, the biodiesel yield was high ≈92%.

Figure 10(e) shows the interactive effect of mixing rate and reaction time, at a constant 6 : 1 M : O and MS-CaO concentration 3 wt%. It is obvious that increasing the mixing rate and reaction time increased the yield to a certain extent recording ≈92% at 200–250 rpm and 75–93 min. Further increment of mixing rate decreased the yield recording ≈81% at 400 rpm and 30 min and 84% at 400 rpm and 120 min.

Figure 10(f) presents the interaction between reaction time and catalyst concentration. The increases of both factors decrease the biodiesel yield, recording maximum yield ≈92% at 3–4.5 wt% and 75–93 min.

From the RSM analysis it can be concluded that the maximum biodiesel production 89−92% using CaO prepared from mollusks shells MS-CaO can be achieved over a wide range of experimental parameters, at 6 : 1–6.5 M : O molar ratio, 3–4.5 wt% catalyst concentration, 75–93 min reaction time, and mixing rate of 200–250 rpm, at 60°C.

Figure 11(a) illustrates the effect of M : O molar ratio and CS-CaO catalyst loading on biodiesel production at a constant reaction time of 120 min and mixing rate 400 rpm. At low M : O ratio the increase of catalyst concentration decreased the biodiesel yield, recording ≈79.6% yield at 6 : 1 M : O and 9 wt% catalyst concentration. This can be explained by the effect of mass transfer limitation with the presence of excess solid catalyst. But at high M : O ratio, the biodiesel yield increased with increase of catalyst loading, recording maximum biodiesel yield of approximately 97% at ≈10.5 : 1–12 : 1 M : O and 7.5–9% catalyst loading (w:w). According to Son et al. [20], a molar ratio of methanol to oil M : O higher than the stoichiometric ratio has generally been adopted for biodiesel production, as methanol would promote the formation of methoxy anions on the catalyst surface, leading to a shift in the equilibrium in the forward direction, thus increasing the biodiesel yield.

Figure 11(b) represents the effect of M : O molar ratio and reaction time on biodiesel production at a constant catalyst concentration 3 wt% and mixing rate 400 rpm. It seems that at low and high M : O, the increase in reaction time decreases the yield, recording ≈80% at 6 : 1 M : O and 120 min and ≈93% at 12 : 1 M : O and 120 min but ≈97% at 12 : 1 M : O and 30 min.

Figure 11(c) represents the effect of M : O and mixing rate on biodiesel production at a constant reaction time 120 min and catalyst loading 3%. The 3D plot shows that, at low and high M : O, the increase in mixing rate decreases the yield, recording ≈97% at 12 : 1 M : O and 200 rpm but ≈93% and 80% at 400 rpm using 12 : 1 and 6 : 1 M : O, respectively.

Figure 11(d) shows that there is a negative interactive effect between these two factors, where at low mixing rate, increasing the catalyst concentration increased the biodiesel yield, recording 80 and 97% at 200 rpm using 3 and 9 wt% catalyst, respectively. But, at high mixing rate 400 rpm, the opposite occurred; increasing the catalyst loading decreased the yield, recording ≈93 and 97% at 9 and 3 wt% catalyst loading, respectively. The separation phase between hydrophilic methanol and hydrophobic oil and between the two liquid reactants and solid catalyst is generally known to be major problem, lowering the biodiesel yield in heterogeneous process. Thus, increasing the stirring rate would increase the mixing of the reactants, which would consequently promote the transesterification process, as the mass transfer of the reactants to the catalyst surface can be enhanced. Further increase in the stirring velocity would lead to increase in turbulence in the reaction mixture, which would lower the biodiesel yield. Also the increase of catalyst concentration would increase the active sites of the solid catalyst available for the transesterification process.

Figure 11(e) shows the interactive effect of mixing rate and reaction time. It is obvious that increasing the mixing rate to a certain limit

200–300 rpm, with increase of reaction time 30–50 min, yields larger amount of biodiesel, recording its maximum ≈97%. Further increase of mixing rate slightly decreased the biodiesel yield, recording ≈93% at 400 rpm and 120 min and ≈89% at 400 rpm and 30 min.

Figure 11(f) presents the negative interaction between reaction time and catalyst loading. The increase of catalyst loading and reaction time decreased the biodiesel yield to reach ≈80% at 9 wt% catalyst concentration and 120 min. The maximum biodiesel yield ≈97% occurred at 9% catalyst concentrations and 30 min reaction time and ≈93% at 3% catalyst concentrations and 120 min reaction time. According to Zhang et al. [21], the reason for the reduction in biodiesel yield with excess catalyst may be attributed to increase in viscosity of reaction mixture which might result in mass transfer limitation.

From the RSM analysis it can be concluded that the maximum biodiesel production using CaO prepared from crabs shells CS-CaO can be achieved at 10.5 : 1–12 : 1 M : O, 7.5–9% catalyst concentration, 30–50 min reaction time, and mixing rate of 200–300 rpm, at 60°C.

# Optimization of the Transesterification Process

The optimization process was carried out to determine the optimum value of biodiesel production efficiency, using the Design Expert 6.0.7 software. According to the software optimization step, the desired goal for each operational condition (A M : O, B catalyst concentration, reaction time, and  mixing rate) was chosen "within" the studied range. The responses (% biodiesel yield) were defined as maximum to achieve the highest performance. The program combines the individual desirability into a single number and then searches to optimize this function based on the response goal. Accordingly, the optimum working conditions and respective biodiesel production were established. The maximum predicted biodiesel yields 94 and 100% were found to be achieved at 6 : 1 and 12 : 1 M : O, 4.5 and 7.3 catalyst wt%, 82 and 30 min reaction time, and 220 and 214 rpm mixing rate, in case of using biocatalyst prepared from mollusks and crabs shells, respectively. The desirability function value was found to be 1.000 for these optimum conditions. An additional experiment was then performed to confirm the optimum results. The laboratory experiment agrees well with the predicted response values ≈96% and 98%, respectively. That indicates the process optimization based on D-optimal design of experiments was capable and reliable to optimize biodiesel production from waste cooking oil using biocatalyst prepared from MS and CS. The lower required amounts of methanol and catalyst concentrations in case of transesterification using MS-CaO relative to those using CS-CaO might be attributed to the recorded higher basicity of MS-CaO relative to that of CS-CaO.

Boey et al. [2] reported the optimum conditions for production of biodiesel ($\approx$98%) from palm olein oil using CaO prepared from waste mud crab shells by calcination above 700°C for 2 h to be 0.5 : 1 M : O g : g, 5 wt% catalyst concentration, 2.5 h, 500 rpm, and 65°C.

Correia et al. [22] reported the production of 83.1% biodiesel yield from sunflower oil using CaO prepared from waste crabs shells calcinated at 900°C for 2 h, using the following conditions: 6 : 1 M : O molar ratio, 3 wt% catalyst concentration, 4 h, 1000 rpm, and 60°C.

Viriya-empikul et al. [1] reported 92% biodiesel yield from palm olein oil using CaO prepared from waste mollusks shells by calcination at 800°C for 2 h to be 18 : 1 M : O molar ratio, 10 wt% catalyst concentration, 2 h, and 60°C.

# Comparing the Efficiency of the Prepared Catalyst with Chemical CaO and Novozym 435

The activity of the prepared MS-CaO and CS-CaO, using the obtained optimum conditions for production of biodiesel, was compared with that of commercially available most effective heterogeneous basic catalysts CaO and Novozym 435.

At optimum conditions of MS-CaO, 6 : 1 M : O, 4.5% catalyst wt%, 82 min, 220 rpm, and 60°C, the achieved biodiesel yield was $\approx$95, 83, and 40% using MS-CaO, chemical CaO, and Novozym 435, respectively. While applying the optimum operating conditions for CS-CaO, 12 : 1 M : O, 7.3 catalyst wt%, 30 min, 214 rpm, and 60°C, the biodiesel yield recorded $\approx$97, 78, and 35% using CS-CaO, chemical CaO, and Novozym 435, respectively. That indicates the higher efficiency of the produced MS-CaO and CS-CaO for production of biodiesel in comparison with the commercial chemical CaO and the most widely used enzyme Novozym 435.

# Physicochemical Characterization of the Produced Biodiesel Using CaO Prepared from MS and CS

The produced biodiesel using the obtained optimum transesterification conditions of each of the prepared biocatalysts was evaluated on the basis of its fuel properties compared to Egyptian petro-diesel and international biodiesel standards as shown in Table 6. All the properties of the produced biodiesel are completely acceptable and meet most of the specifications. So it can be ranked as a realistic fuel and as an alternative to petro-diesel. This recommends the applicability of waste mollusks and crabs shells for production of cheap catalyst to produce biodiesel in a cost effective process.

**Table 6**: Physicochemical properties of the produced biodiesel using CaO prepared from waste MS and CS compared to international standards of bio-diesel and Egyptian petro-diesel standard specifications

| Test | Unit | Produced biodiesel | | Egyptian perto-diesel standards | Biodiesel EN14214 | Biodiesel D-6751 |
|---|---|---|---|---|---|---|
| | | Mollusks shells CaO | Crabs shells CaO | | | |
| Density at 15.56°C | g/cm³ | 0.8914 | 0.9003 | 0.82–0.87 | 0.86–0.9 | — |
| Specific gravity | | 0.8924 | 0.9011 | — | — | — |
| API | | 27.06 | 25.53 | — | — | — |
| Kinematic viscosity at 40°C | cSt | 5.5 | 7.6 | 1.6–7 | 3.5–5 | 1.9–6 |
| Pour point | °C | −2 | −2 | 4.5 | — | — |
| Cloud point | °C | 1 | 1 | — | — | — |
| Total acid number | mg KOH/g | 0.6 | 0.7 | Nil | <0.5 | <0.8 |
| Total S | wt% | 0.003 | 0.006 | <1.2 | <0.01 | <0.05 |
| Water content | ppm | 312 | 300 | 1500 | <500 | <500 |
| Flash point | °C | 148 | 143 | >55 | >101 | >130 |
| Calorific value | MJ/Kg | 38.15 | 35.55 | >44.3 | 32.9 | — |
| Iodine number | mg I₂/100 g | 105 | 108 | — | <120 | — |

The iodine value of the produced biodiesel fuel BDF was in the range of 105–108 mg $I_2$/100 g oil. Iodine value is a measure of unsaturation degree. The degree of unsaturation greatly influences fuel oxidation tendency. According to EN 14214, methyl esters used as diesel fuel must have an iodine value less than 120 mg $I_2$/100 g sample.

The acid value measures the content of free acids in the sample, which has influence on fuel aging. The acid value of produced biodiesel was 0.6 and 0.7 mg KOH/g, with average lowering of ≈80 and 76.67% from the used WCO, indicating better transesterification efficiency, using CaO prepared from MS than that prepared from CS. The TAN of the produced biodiesel is relatively high, but within the ASTM D6751 biodiesel standards. Candeia et al. [23] reported that the biodiesel with high TAN causes operational problems, such as corrosion and pump plugging, caused by corrosion and deposit formation.

Felizardo et al. [24] reported that density at 15°C and kinematic viscosity at 40°C are important properties, mainly in airless combustion systems because they influence the efficiency of atomization of the fuel, flow, and distribution.

The density of the produced biodiesel fuel using MS-CaO and CS-CaO recorded 0.8914 and 0.9003 g/cm³ compared to that of petro-diesel 0.8421 g/cm³. Fuel with high paraffincity has high specific gravity and low API. The produced biodiesel was characterized by higher specific gravity (0.8924 and 0.9011, resp.) and lower API value (27.06 and 25.53, resp.) compared to those of the Egyptian petro-diesel sample (0.8428 and 36.39, resp.). Therefore, volumetrically, biodiesel delivers a slightly greater amount of fuel [19].

The viscosity of the produced biodiesel recorded 5.5 and 7.6 cSt with a remarkable decrease from that of WCO of ≈89 and 84.8% through transesterification using CaO prepared from MS and CS, respectively.

The better recorded TAN, density, and viscosity of the produced biodiesel using MS-CaO would indicate better transesterification efficiency, using CaO prepared from MS than that prepared from CS. This might be attributed to the recorded higher basicity of MS-CaO relative to that of CS-CaO.

The produced biodiesel has acceptable cold flow properties pour point PP −2°C and cloud point 1°C and is characterized by lower CV

(38.15–35.55 MJ/kg, resp.) relevant to that of the Egyptian petro-diesel sample (45.49 MJ/kg).

The water content of the produced biodiesel is higher than that of the Egyptian petro-diesel sample recording 312–300 and 84 ppm, respectively. But it is within the recommendable biodiesel standards, <500 ppm.

The produced biodiesel has three major advantages; it is ultralow sulfur biofuel with sulfur content of 0.003–0.006%, while petro-diesel has 0.2% sulfur. So it meets the aim of petroleum industry for ultralow sulfur diesel fuel and the biodiesel combustion will not produce large amount of sulfur oxides which lead to corrosion of the engine parts and environmental pollution. The produced biodiesel has a higher FP 148–143°C, compared to 63°C for petro-diesel. So biodiesel is much less flammable fuel than petro-diesel and hence it is much safer in handling, storage, and transport. In addition, the viscosity of the produced biodiesel 5.5–7.6 cSt is competitive to regular Egyptian standards for petro-diesel 1.6–7 cSt. Hence, no hardware modifications are required for handling the produced BDF in the existing engine.

The better qualifications of the produced biodiesel, using CaO prepared from MS than those of biodiesel produced using CaO prepared from CS, might be attributed to the higher basicity of MS-CaO than that of CS-CaO which would positively impact the transesterification reaction.

# CONCLUSIONS

In the viewpoint of preparation time, energy consumption, cost of catalyst, and high quality biodiesel yield, the CaO catalyst prepared from waste mollusks and crabs shells can be recommended for biodiesel production from waste cooking oil collected from seafood restaurants. This would have a triple positive impact on environment, economic, and energy sectors.

Further work is undertaken now in EPR-Biotechnology laboratory concerning the kinetics and mechanism of the transesterification process, the reusability, and stability of the prepared biocatalysts compared to those of the chemical CaO and Novozym 435, and their effects on the biodiesel yield, its purity, and quality.

# CONFLICT OF INTERESTS

The authors declare that there is no conflict of interests regarding the publication of this paper.

# ACKNOWLEDGMENTS

The authors are grateful to Dr. Samiha F. Deriase, Professor of Chemical Engineering in Egyptian Petroleum Research Institute, for her help in D-optimal design of experiments.

# REFERENCES

1. N. Viriya-empikul, P. Krasae, W. Nualpaeng, B. Yoosuk, and K. Faungnawakij, "Biodiesel production over Ca-based solid catalysts derived from industrial wastes," Fuel, vol. 92, no. 1, pp. 239–244, 2012.

2. P.-L. Boey, G. P. Maniam, and S. A. Hamid, "Biodiesel production via transesterification of palm olein using waste mud crab (Scylla serrata) shell as a heterogeneous catalyst," Bioresource Technology, vol. 100, no. 24, pp. 6362–6368, 2009

3. N. S. El-Gendy, S. F. Deriase, and A. Hamdy, "The optimization of biodiesel production from waste frying corn oil using snails shells as a catalyst," Energy Sources, Part A: Recovery, Utilization and Environmental Effects, vol. 36, no. 6, pp. 623–637, 2014

4. M. Kouzu and J.-S. Hidaka, "Transesterification of vegetable oil into biodiesel catalyzed by CaO: a review," Fuel, vol. 93, pp. 1–12, 2012

5. N. El-Gendy and H. Madian, Ligno-Cellulosic Biomass for Production of Bio-Energy in Egypt, Lambert Academic, Saarbrücken, Germany, 2013.

6. J. Boro, A. J. Thakur, and D. Deka, "Solid oxide derived from waste shells of Turbonilla striatula as a renewable catalyst for biodiesel production," Fuel Processing Technology, vol. 92, no. 10, pp. 2061–2067, 2011

7. ASTM Standards Methods, Annual Book of ASTM Standards. Petroleum Products and Lubricants (I–III), American Society for Testing and Materials, West Conshohocken, Pa, USA, 1991.

8. JUS EN 14214, Automotive Fuels. Fatty Acid Methyl Esters (FAME) for Diesel Engines-Requirements and Test Methods, Standardization Institute, Belgrade, Serbia, 2004.

9. ASTM Standard D6751, Standard Specification for Bio-Diesel Fuel (B100) Blend Stock for Distillate Fuels. West Conshohocken, PA: ASTM. Automotive Fuels-Fatty-Acid Methyl Esters (FAME) for Diesel Engines—Requirements and Test Methods, Beuth-Verlag, Berlin, Germany, 2008.

10. N. B. Singh and N. P. Singh, "Formation of CaO from thermal decomposition of calcium carbonate in the presence of carboxylic acids," Journal of Thermal Analysis and Calorimetry, vol. 89, no. 1, pp. 159–162, 2007

11. B. Engin, H. Demirta , and M. Eken, "Temperature effects on egg shells investigated by XRD, IR and ESR techniques," Radiation Physics and Chemistry, vol. 75, no. 2, pp. 268–277, 2006. View at Publisher

12. N. Viriya-empikul, P. Krasae, B. Puttasawat, B. Yoosuk, N. Chollacoop, and K. Faungnawakij, "Waste shells of mollusk and egg as biodiesel production catalysts," Bioresource Technology, vol. 101, no. 10, pp. 3765–3767, 2010

13. B. Yoosuk, P. Udomsap, B. Puttasawat, and P. Krasae, "Improving transesterification acitvity of CaO with hydration technique," Bioresource Technology, vol. 101, no. 10, pp. 3784–3786, 2010

14. R. H. Borgwardt, "Sintering of nascent calcium oxide," Chemical Engineering Science, vol. 44, no. 1, pp. 53–60, 1989

15. W. G. Liu, N. W. L. Low, B. Feng, G. Wang, and J. C. Diniz da Costa, "Calcium precursors for the production of CaO sorbents for multicycle $CO_2$ capture," Environmental Science and Technology, vol. 44, no. 2, pp. 841–847, 2010

16. A. Birla, B. Singh, S. N. Upadhyay, and Y. C. Sharma, "Kinetics studies of synthesis of biodiesel from waste frying oil using a heterogeneous catalyst derived from snail shell," Bioresource Technology, vol. 106, pp. 95–100, 2012

17. W. Roschat, M. Kacha, B. Yoosuk, T. Sudyoadsuk, and V. Promarak, "Biodiesel production based on heterogeneous process catalyzed by solid waste coral fragment," Fuel, vol. 98, pp. 194–202, 2012

18. S. Hu, Y. Wang, and H. Han, "Utilization of waste freshwater mussel shell as an economic catalyst for biodiesel production," Biomass and Bioenergy, vol. 35, no. 8, pp. 3627–3635, 2011

19. B. K. Sharma, U. Rashid, F. Anwar, and S. Z. Erhan, "Lubricant properties of Moringa oil using thermal and tribological techniques," Journal of Thermal Analysis and Calorimetry, vol. 96, no. 3, pp. 999–1008, 2009

20. S. M. Son, K. Kusakabe, and G. Guan, "Biodiesel synthesis and properties from sunflower and waste cooking oils using CaO catalyst under reflux conditions," Journal of Applied Sciences, vol. 10, no. 24, pp. 3191–3198, 2010

21. L. Zhang, B. Sheng, Z. Xin, Q. Liu, and S. Sun, "Kinetics of transesterification of palm oil and dimethyl carbonate for biodiesel production at the catalysis of heterogeneous base catalyst," Bioresource Technology, vol. 101, no. 21, pp. 8144–8150, 2010.

22. L. M. Correia, R. M. A. Saboya, N. de Sousa Campelo et al., "Characterization of calcium oxide catalysts from natural sources and their application in the transesterification of sunflower oil," Bioresource Technology, vol. 151, pp. 207–213, 2014

23. R. A. Candeia, M. C. D. Silva, J. R. Carvalho Filho et al., "Influence of soybean biodiesel content on basic properties of biodiesel-diesel blends," Fuel, vol. 88, no. 4, pp. 738–743, 2009

24. P. Felizardo, M. J. Neiva Correia, I. Raposo, J. F. Mendes, R. Berkemeier, and J. M. Bordado, "Production of biodiesel from waste frying oils," Waste Management, vol. 26, no. 5, pp. 487–494, 2006. View at Publisher

# A Review of Enzymatic Transesterification of Microalgal Oil-Based Biodiesel Using Supercritical Technology

Hanifa Taher, [1] Sulaiman Al-Zuhair, [1] Ali H. Al-Marzouqi, [1] Yousef Haik, [2] and Mohammed M. Farid[3]

[1]Chemical and Petroleum Engineering Department, UAE University, Al-Ain 17555, United Arab Emirates

[2]Mechanical Engineering Department, UAE University, Al-Ain 17555, United Arab Emirates

[3]Chemical and Materials Engineering Department, University of Auckland, 1142 Auckland, New Zealand

## ABSTRACT

Biodiesel is considered a promising replacement to petroleum-derived diesel. Using oils extracted from agricultural crops competes with their use as food and cannot realistically satisfy the global demand of diesel-fuel requirements. On the other hand, microalgae, which have a much higher oil yield per hectare, compared to oil crops, appear to be a source that has the potential to completely replace fossil diesel. Microalgae oil extraction is a major step in the overall biodiesel production process. Recently, supercritical carbon dioxide ($SC-CO_2$) has been proposed to replace conventional solvent extraction techniques because it is nontoxic, nonhazardous, chemically stable, and inexpensive. It uses environmentally acceptable solvent, which can easily be separated from the products. In addition, the use of $SC-CO_2$ as a reaction media has also been proposed to eliminate the inhibition limitations that encounter biodiesel production reaction using immobilized enzyme as a catalyst. Furthermore, using $SC-CO_2$ allows easy separation of the product. In this paper, conventional biodiesel production with first generation feedstock, using chemical catalysts and solvent-extraction, is compared to new technologies with an emphasis on using microalgae, immobilized lipase, and $SC-CO_2$ as an extraction solvent and reaction media.

# INTRODUCTION

Continuous exploration and consumption of fossil fuels have led to a decline in worldwide oil reserves. As the world energy demand is continuously increasing, the most sufficient way to meet the growing demand is by finding alternative fuels. From the point of environment protection, finding alternative fuels that are sustainable and environment friendly is essential.

More than a century ago, Rudolf Diesel tested the suitability of using vegetable oils as fuel in his engine [1, 2]. In the 1930s and 1940s, vegetable oils were used as a diesel fuel for emergency situations. At that time, vegetable oil fuels were not competitive because they were more expensive than petroleum fuels, and therefore the idea was abandoned. With the worries about petroleum fuel availability and latest increases in petroleum prices, using vegetable oils in diesel engines has regained attention.

A number of studies have shown that triglycerides (TGs) hold promise as alternative diesel engine fuels [2, 3]. This has an advantage of being available, renewable with higher cetane number, and biodegradable [4–6]. However, the main disadvantage of oils is their high viscosity and low volatility [2, 7, and 8]. Therefore, direct use of TGs is generally unacceptable and not practical since it causes engine coking, carbon depositing and gelling of the lubricating oil [8–10]. To overcome these problems, dilution, pyrolysis, cracking, and Tran's esterification of the oil have been suggested [5, 11]. Among all these methods, Tran's esterification has been used widely as a favorable method. Transesterification reaction of TGs, known as alcoholysis, is an important reaction that produces fatty acids alkyl esters (FAAE) [8, 12]. It was reported that replacing petroleum diesel with FAAE results in a reduction of unburned hydrocarbons, carbon monoxide (CO), and particular matter (PM) formation [13, 14]. Several methods of transesterification using alkali catalysts [9, 14–18], acid catalysts [17, 19–23], and enzyme lipase in presence and absence of solvents have been reported [24–29]. Most of the commercial biodiesel processes require the use of a catalyst, which requires a recovery unit to separate reaction products and remove the catalyst. These disadvantages of using catalyst could be eliminated by carrying out noncatalytic reaction. Sake and Kusdiana [30] developed a method using supercritical methanol (SCM) where triglycerides fatty acids were converted to methyl esters without using any catalyst. Sake and Kusdiana [30] and Madras et al. [31] reported the advantage of using supercritical alcohols (SCA), especially methanol, whereas a process requires short reaction time and no need for reaction product separation from the solvent. However, this process is energy intensive as it is carried out at the supercritical conditions of methanol. Nevertheless, based on van Kasteren and Nisworo [32] economic assessment, this process appears to be feasible.

Enzymatic biodiesel approach showed promising results due to their high selectivity and mild operative conditions. Enzymatic transesterification reaction is similar to conventional transesterification, except that they are catalyzed by a variety of biological catalysts rather than chemical catalysts. In contrast to conventional processes, biocatalysts can transesterify TGs with a high free fatty acid (FFA) content [33]. Lipase-catalyzed transesterification of TG has been investigated by several investigators [33–37]. One common drawback with the use of enzyme-based processes is the high cost of the enzyme compared to conventional chemical catalysts; therefore, their recycle is required, which is possible through enzyme immobilization.

Immobilization of enzymes has generally been used to attain reusable enzyme with lower production cost [25, 38, 39]. Thus, immobilized form of lipase has been used in most of transesterification processes [25, 36,40]. Besides enzyme reusability, other advantages of using immobilized lipase as a catalyst are enhanced activity and stability [41, 42].

Several researches have been carried out to produce biodiesel in solvent systems. Presently, industries are facing problems in using conventional solvents due to environmental worries. In the last couple of decades, enzyme-catalyzed reactions in supercritical carbon dioxide (SC-$CO_2$) has been studied. Previously, most of the studies were investigating the feasibility of using biocatalyst in SC-$CO_2$, whereas recent studies are focusing on obtaining good yield and conversions.

Vegetable oils consist of TG of straight chains of fatty acids. With the high cost of biodiesel produced from vegetable oils, researchers are looking for low-cost feedstocks. For that waste oils, cooking oils and fats from animal sources were proposed. The main drawback of using animal fats is their high melting points, which may require the use of organic solvents. However, organic solvent use requires a solvent recovery unit and energy needed for its separation. To overcome this, supercritical fluids (SCFs) were introduced.

During the past decades, SCFs have been investigated as alternative solvents for reactions rather than using conventional solvents. Among all supercritical fluids, SC-$CO_2$ is the most appropriate choice as a consequence of its availability. In general, $CO_2$ is nontoxic, nonflammable, environmentally friendly, and recyclable fluid [43]. Thus, reactions in SC-$CO_2$ media become the preferable route for chemical synthesis.

Conventionally, biodiesel is produced from vegetable oils, animal fats, and waste cooking oils [7, 30, 31, 44–46]. However, these feedstocks are inefficient and unsustainable [47]. Furthermore, using vegetable oil as a fuel source competes with its use as food and proposes for land development in order not to compete with food and land. On the other hand, animal fat cannot be considered as a continuous supply of feed stock [48]. Thus, biodiesel production using these feedstocks, realistically, cannot replace all world biodiesel requirements.

In contrast, microalgae have been recognized as a promising alternative source for biodiesel production. They are a group of organisms that can grow photosynthetically and accumulate large amounts of lipids [49, 50]. According to Sheehan et al. [50], if microalgal oil production could be scaled up, less than 6 million hectares would be required to meet current fuel demands.

Considering the above facts, this paper provides an overview on biodiesel production from microalgae with a particular emphasis on the use of microalgae as a promise feedstock, lipase as promise catalyst, and SC-$CO_2$ as a promise extraction solvent and reaction media.

# BIODIESEL

Biodiesel has arisen as a possible alternative for petroleum diesel because of the similarities that biodiesel has with petroleum diesel [39, 51]. Biodiesel fuel has many advantages over petroleum fuel such as being nontoxic, biodegradable, renewable, and do not contribute to net accumulation of the greenhouse gases [52,53]. Also, biodiesel has lower sulfur and aromatic content, higher cetane number, and flash point than petroleum diesel [5, 7, 54–56]. Other benefits of biodiesel include increased lubricity and lower emissions of certain harmful exhaust gases in comparison to petroleum diesel fuel [55].

Comparing petroleum diesel fuel to biodiesel, Schumacher et al. [14] reported that biodiesel results in a 45% reduction in total hydrocarbon emissions, 47% reduction in CO emissions, and 66% reduction in PM emissions, whereas, Demirbas [55] reported a 42% reduction in CO and 55% in PM emissions relative to standard diesel fuel. These effects are generally attributed to the higher cetane number and oxygen content of biodiesel fuel. Although the biodiesel environmental considerations are very positive, biodiesel increases nitrogen oxides () emissions. However, reports show that reductions in emissions are possible with some modifications in combustion temperatures and injection timing [57, 58].

As mentioned earlier, direct use of vegetable oil has several negative aspects, such as their high viscosity and low volatility, which lead to incomplete combustion in diesel engines, therefore, carbon deposition [5, 9, 59, and 60]. However, the direct use of vegetable oils as biodiesel may be possible by mixing them with conventional diesel in an appropriate ratio, but this mixing will be impractical for long-term uses in the engine due to the high viscosity, low stability, acid composition, and FFA content [4, 61, and 62]. Therefore, considerable efforts have been made to develop vegetable oil derivatives that have properties near those of the petroleum-based diesel fuels.

Pyrolysis (cracking), microemulsion, and transesterification are the possible methods to minimize problems associated with feedstock use [5, 8, and 11]. The first two methods are costly and yield low quality biodiesel, whereas the latter, transesterification, is the most common method to transform oil into biodiesel, which is the focus of this paper.

# Transesterification

Transesterification is the common method used to transform TG into biodiesel. This consists of the reaction between TG and an acyl-acceptor [11, 63]. Carboxylic acids, alcohols, or another ester can be used as acyl-acceptor. Transesterification produces glycerol when alcohol is used as acyl-acceptor or triacylglycerol when ester is used [8, 56, 60, 62, and 64]. Transesterification process using a catalyst is called catalytic transesterification process, whereas that without catalyst is called noncatalytic transesterification process [8, 10, 65, and 66]. Moreover, catalytic process is divided into two types: homogenous and heterogeneous processes depending on the catalyst used.

Transesterification is a chemical process of transforming large and branched TG into smaller and straight chain molecules, which is similar in size to the molecules of the species present in diesel fuel [67, 68]. Stoichiometrically, for each mole of TG three moles of alcohol are required. But in general, a higher molar ratio of alcohol is used in order to achieve maximum biodiesel production. This molar ratio depends on the type of used feedstock, type of catalyst, temperature, and reaction time. Methanol, ethanol, and propanol are the most commonly used alcohols. In fact, biodiesel yield is independent of the type of the alcohol used and the alcohol selection depends on cost [60]. In transesterification, ester bonds are broken first then followed by hydroxyl bond, whereas in esterification hydroxyl bonds are broken before ester bonds resulting in glycerol as byproduct in transesterification and water in esterification [54].

Transesterification can be carried in a number of ways, using different catalytic processes. For example, it can be carried out using alcohol and alkali catalyst, acid catalyst, and biocatalyst or using alcohols in their supercritical state [39, 69]. Overall, transesterification is a sequence of three reactions; TG is first converted to a diacylglycerol (DG) and one fatty acid ester, then the DG is converted to monoacylglycerol (MG) giving an additional fatty acid ester, and finally the MG is converted to glycerol giving the last fatty acid ester.

Catalyst promotes hydrolysis of the TGs to produce fatty acids and glycerol, with the last being a byproduct. By the end of the transesterification, produced biodiesel and glycerol have to be purified in order to remove the catalyst, which requires a separation step by washing with distilled water for several times. It is well understood that catalyst selection is an important criterion.

# Chemical Catalytic Transesterification

## *Alkaline Catalyst Transesterification*

A base catalyst is a chemical with a pH value greater than 7. It has the ability to give extra electrons. Sodium hydroxide (NaOH), potassium hydroxide (KOH), and sodium methoxide ($CH_3ONa$) are the most common homogeneous base catalysts employed during alkaline transesterification [11, 51, 60]. The base catalyzed process is the most commonly used because of its relative ease. It can be performed at low temperature and pressure and yields high conversion (98%) within a short time [5]. Most important limitation of the base catalysis method is the process sensitivity to both FFA and water contents. It works perfectly when the FFA and moisture contents are less than certain limits, usually below 0.5 wt% for FFA [70, 71]. In case of TGs where FFA contents exceed this limit, pre-treatment step is required. The presence of FFA promotes soap formation, which consumes the catalyst, lowers the yield, and more importantly results in difficult downstream byproducts separation and product purification [7, 8]. About 60–90% of biodiesel cost comes from the high cost of the raw material [7]. In addition, alkali catalyst needs effluent treatment. Most of the base catalyzed reactions were carried out at temperatures close to the alcohol boiling point with alcohol to oil molar ratio of 6 : 1. Akoh et al. [9] stated that to increase biodiesel yield, a stoichiometric excess of substrates (6 : 1 molar ratio of methanol to oil) is favored. Homogeneous catalysts have been used industrially for biodiesel production where produced biodiesel and glycerol have to be purified to remove the catalyst. This purification process requires large quantities of water and energy. Thus, heterogeneous catalysts have been suggested to overcome this drawback. Heterogeneous catalysts can be separated easily from the system at the end by filtration and could be reused [60, 72]. Alkaline earth oxides [73], zeolites [3], calcined hydrotalcites [18, 74], and Magnesium and Calcium oxides [16, 72] have been suggested as heterogeneous catalysts and showed good results. However, the high cost of the purified feedstock remains the main problem facing the alkaline catalyzed process.

Acid Catalyst TransesterificationThe reaction of TGs and alcohol may also be catalyzed with an acid instead of a base. Most commonly used acids are strong acids like sulphuric, sulphonic, phosphoric, and hydrochloric acids [5, 8, 10]. Acid-catalyzed transesterification processes are not as popular as the base-catalyzed processes, mainly because strong acids are corrosive and the processes are too slow. Several reactions may be required in order to achieve high conversion. It has been stated that acid-catalyzed reaction may be 4000 times slower than the base catalyst process [7, 9, 54, 66]. Above that, it requires high amount of alcohol and higher concentration of catalyst. Akoh et al. [9] stated that a molar methanol:oil ratio of 30:1 in a range of 55–80°C with 0.5 to 1 mol% catalyst concentration is required to achieve 99% conversion in 50 h. On the other hand, acid-catalyzed processes offer an important advantage for being independent of feedstock FFA content. That is because feedstock FFA is not converted to soap using this kind of catalysts, and hence biodiesel can be produced from low cost feedstock [6, 54]. As mentioned before, feedstock of high FFA content requires a pretreatment step if a base catalyst is to be used. This pretreatment step can be achieved using acid catalysis and methanol, where the FFA is esterified to biodiesel. When equilibrium is reached, the acid catalyst and produced water are removed from the reaction vessel by centrifugation [11]. This is followed by adding fresh methanol and base catalyst to the oil in order to catalyze the transesterification reaction. Heterogeneous acid catalysts have been also used. This is important to avoid problems associated with homogeneous catalysts. Sulphated tin oxide has been used as superacid catalysts to transesterified waste cooking oil [22]. Sulphated zirconia was also used as catalysts in the alcoholysis of soybean oil and in the esterification of oleic acid [21].

Heteropolyacid was used to transesterify yellow horn oil [75]. Anion and cation exchange resins were used for triolein transesterification reactions with ethanol to produce ethyl oleate [76].

## Noncatalytic Transesterification

Although catalysts play a great role in reducing transesterification time, their presence promotes complications of final product purification. This results in increased production process cost.

To avoid catalyst drawbacks, supercritical alcohol (SCA) transesterification process was suggested [13, 51,77]. SCA transesterification process is a catalyst free process, which provides high conversion of oil to ester in a short time. Tan et al. [78] compared SCM transesterification with conventional catalytic methods. They reported that conventional catalyst required 1 hr to convert palm oil to biodiesel, whereas SCM required only 20 min. As a result of catalyst absence, purification of the products of the transesterification reaction is much simpler and environmentally friendly compared to the previously mentioned processes.

In 2001, Saka and Kusdiana [30] conducted a research on biodiesel production from vegetable oils without any aid of catalysts. The oil-methanol mixture was heated above the supercritical temperature. Biodiesel was removed from the reaction mixture, and the excess methanol was removed by evaporation for 20 min at a temperature of 90°C. It was reported that 95% conversion was achieved in the first 4 min of reaction with optimum process parameters of alcohol : oil molar ratio of 42 : 1, pressure of 430 bar, and reaction temperature of 350°C. After one year (2002), Demirba [77] studied transesterification of six different vegetable oils in supercritical methanol and reported that increasing reaction temperature to supercritical condition had favorable influence on ester conversion.

Compared to catalytic reactions, SCM reactions are fast and can achieve high conversions in a very short time. However, the reaction requires higher temperatures, pressures, and alcohol to oil molar ratio in comparison to catalytic transesterification, which result in high production cost [67, 68].

It is clearly shown that the three transesterification processes presented have several drawbacks. They are energy intensive, recovery of byproduct is difficult, catalysts have to be removed, and waste treatment is required. To overcome these problems, enzymes have been proposed [7, 54, 79]. Most important advantage of using enzymes is their ability to convert FFA contained in the fat or oil to methyl esters completely. Additionally, glycerol, byproduct, can be easily recovered [26, 51, 80].

## Enzymatic Transesterification

There is a great interest on using biocatalysts to catalyze TG transformation to biodiesel, which has the advantage of having low operating conditions and high product purity. Enzymatic transesterification can be carried out at 35 to 45°C [41, 42, 81]. Contrary to chemical catalysts, enzymes do not form soaps and catalyze esterification of FFA and TG in one step without any need of the washing step. On the other hand, the major disadvantages of the enzymatic transesterification are its slower reaction rate and possible enzyme inactivation by methanol [27, 62, 82]. Lipase is an enzyme capable of catalyzing methanolysis reactions. It can be obtained from microorganisms such as bacteria and fungi. Lipases from Mucor miehei, Rhizopus oryzae, Candida antarctica, and Pseudomonas cepacia are the most commonly used enzymes [39, 62]. Lipases belong to a group of hydrolytic enzymes called hydrolases. In biological systems, lipases hydrolyze TGs to fatty acids and glycerol [66]. They work in mild conditions and have an ability to work with TGs from different origins. They have the ability to catalyze transesterification of both TGs and FFAs to give esters.

Extracellular and intracellular lipases are the major biocatalyst [5, 56]. Extracellular lipases refer to the recovered enzymes from the microorganism which is then purified, whereas intracellular lipases, the enzyme remains inside the producing cell walls [62]. In term of regioselectivity, lipases have been divided into three types [81]: (i)sn-1,3-specific: hydrolyze ester bonds in positions $R_1$ or $R_3$ of TG, (ii)sn-2-specific: hydrolyze ester bond in position $R_2$ of TG, (iii)nonspecific: do not distinguish between positions of ester.

Fjerbaek et al. [39] stated that for biodiesel production from TG, lipases should be nonstereospecific where all TG, DG, and MG can be converted to fatty acids methyl esters (FAME). In addition, they should also be able to catalyze FFA esterification.

Despite the lipases advantages over acid and base catalysts, lipases are costly which limit their industrial use [60, 83]. For that reason, reusability of the enzyme by using it in an immobilized form is essential from economic point of view.

Soluble enzyme acts as a solute in that they are dispersed in the solution and can move freely, but at the same time difficult to separate and to handle. One promising approach to overcome this difficulty is to immobilize the enzyme in a way that can be separated later by any simple separation method. Enzyme immobilization is a technique where free movement of the enzyme is restricted and localized to an inert support or carrier. This technique has many advantages, the most important of which is that the immobilized enzyme can be reused [40, 84]. In addition, by immobilization, the operating temperature of the process can be increased [39]. Cao [84] mentioned that an immobilized enzyme has to perform two essential functions, namely, the noncatalytic functions that are designed to aid separation and the catalytic functions that are designed to convert the targeting substrates within a desired time. This is in addition to the fact that the process is environmentally friendly and more sustainable [82].

Enzyme immobilization can be carried out in different ways. It can be classified into chemical and physical methods as shown in Figure 1. In biodiesel enzymatic production, various immobilization techniques have been used. Du et al. [24] used adsorption on macroporous resin, Noureddini et al. [27] worked on hydrophobic sol-gel support by entrapment, and Orçaire et al. [28] worked on silica aerogel by encapsulation.

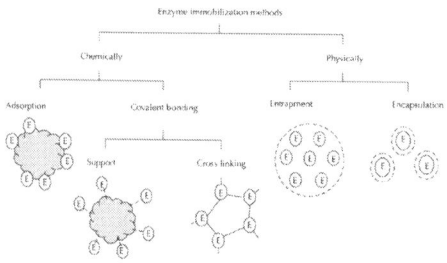

**Figure 1:** Enzyme immobilization methods.

Amongst all possible immobilization methods, physical adsorption has been clearly selected by most researchers due to its ease, the absence of expensive and toxic chemicals, ability to retain the activity, and feasibility of regeneration [54]. But, immobilized enzymes are also subjected to diffusion limitation (internal and external) and inactivation (mostly by methanol) [85]. These problems have been studied and solved by different researchers.

To overcome immobilized lipase inactivation, the addition of an inert solvent has been suggested. However, solvent addition is not highly recommended since this will require using solvent recovery units, which will increase production cost.

Köse and coworkers [86] investigated lipase-catalyzed transesterification of cotton seed oil with methanol in solvent-free medium. Yield of 92% was achieved in the presence of the Novozym 435 in 24 h reaction. This was performed at 50°C, 4:1 alcohol to oil and 30% enzyme loading. In 2007, Royon et al. [87] performed comparable work for cotton seed oil using Novozym 435 at the same condition, but with using tert-butanol as solvent. They noted that tert-butanol dissolved both methanol and glycerol that might inhibit enzyme activity, and a higher conversion of 97% was observed after 24 h of reaction. Similarly, Nelson et al. [33] tested the effect of using solvent on biodiesel production yield. By using Mucor miehei lipase, yield of 94% was obtained in n-hexane system, whereas only 19% yield obtained in a solvent-free system after 8 h reaction with methanol. On the other hand, ethanol produced 65% yield in solvent-free system and 98% in n-hexane system within first 5 h of the reaction. In the same approach, using 80% tert-butanol (based on oil weight) improved biodiesel yield from soybean oil deodorizer distillate with 4% Novozym 435 from 80 to 84%. However, further increase in solvent use decreased the yield, which might result from the dilution effect on reactants [88]. Wang et al. [88] obtained same yield, 84%, when a combination of 2% Novozym 435 and 3% Lipozyme TL were used.

The importance of using solvents has been addressed well in the literature. However, as mentioned earlier solvents can be toxic, flammable, and have to be separated from the ester for reuse. Hence, efforts have been made to offer alternative solvents that are nontoxic and environmentally friendly. Candidate solvents that can replace previously mentioned solvents should have same advantages of dissolving both substrates and reduce excess alcohol inhibition and at the same time avoid the drawbacks of difficult separation of the solvent. In this regard, supercritical fluids (SCF) have been suggested as alternative solvents [89]. Further discussion on the use of SC-$CO_2$ as a reaction medium is found in Section 4.1.2.

# Biodiesel Feedstocks

Biodiesel can be synthesized from a great variety of feedstocks. These feedstocks include most vegetable oils (soybean oil, jatropha oil, rapeseed oil, palm oil, sunflower oil, corn oil, peanut oil, canola oil, and cottonseed oil) and animal fats (tallow and lard). They can also be produced from other sources like waste cooking oil, greases, and oleaginous microorganisms with excess microbial lipid such as microalgae [13].

## *Vegetable Oils*

Since vegetable oil is a feedstock that is available in large quantities, it has been widely used for the conversion to biodiesel. Majority of vegetable oils have been employed for biodiesel production such as soybean oil [44, 90], rapeseed oil [30, 80], canola oil [64], palm oil [45, 54, 91], and sunflower oil [92, 93]. However, producing biodiesel from vegetable oils competes with their use as food and involves additional land use. Also, in industrial scale, biodiesel production requires considerable use of arable lands.

## *Waste Cooking Oils*

Fried oils and fats are usually broken down after a period of use and become unsuitable for further cooking as a result of increasing of FFA content. Once this reached, they are discarded or recycled. This type of feedstocks is of low cost, making them attractive for fuel production [46]. Using waste cooking oil, especially those that cannot be treated, will reduce the environment pollution. Waste cooking oil conversion into biodiesel through the transesterification process reduces their molecular weight to approximately one-third, viscosity by about one-seventh, as well as reducing their flash point and volatility [4, 51]. High oil conversion (>90%) has been reported by many investigators [94–96] in spite of the high FFA contents that range from 5 to 15 wt%.

## Animal Fat

Animal fats are received from cattle, hog, chicken, lamb, and fish. Tallow and animal meats which are not allowed to be used as food can be used as biodiesel feedstock. However, these two sources have discontinuity problem in their supply. It is possible that suddenly a high bulk of material is available followed by a period with no supply like in the case of animal disease [48]. Animal fats are characterized by the high amount of saturated fatty acids (SFA). They are solid at room temperature and cannot be used as fuel in a diesel engine in their original form [97]. Authors of this paper had investigated possibility of biodiesel production from lamb meat fat [98] and tallow [99, 100] as feedstock.

## Oleaginous Microorganisms

As an alternative to vegetable oils and animal fats, oleaginous microorganisms have recently attracted great attention. It has been reported that such microorganisms accumulate oils and have microbial lipid content exceeding 20% [13]. The scope of this paper is on the use of microalgae in biodiesel production.

Using algae as a feedstock has been studied worldwide by several decades. However, for biodiesel production, this was started by an 18-year National Renewable Energy Laboratory (NREL) research project [50]. The potential of using algae for biodiesel production can be seen from their ability to produce large amount of biodiesel and reduce the production cost. Based on algae size, they are classified to macroalgae and microalgae. Macroalgae are large and multicellular, whereas microalgae are small and unicellular. Due to the simple cell structure, microalgae are widely used and have been accepted as promise feedstock. The following section gives more details about microalgae and their potential as feedstock for biodiesel production.

# MICROALGAE

Algae that contain chlorophyll are photosynthetic microorganisms that convert inorganic carbon, such as carbon dioxide, in the presence of light, water and nutrients to algal biomass [101–106]. Majority of algae are living in aquatic (saline or freshwater) environments, whereas some of them can be found in other environments such as snow, desert soils, and hot springs [107]. They can be either autotrophic or heterotrophic. Autotrophic algae require only carbon dioxide, light, and salts to grow, whereas heterotrophic require an organic source of carbon, like glucose, as well as nutrients [105, 108–110]. However, heterotrophic algae are not as efficient as autotrophic algae for oil production [49, 111]. Autotrophic is more favorable as it does not require glucose which is a food source and at the same time fixes $CO_2$, which has positive effect on the environment. Microalgae also can be either phototrophic or chemotrophic. Phototrophic algae use light as an energy source, whereas chemotrophic type use oxidizing compounds [107]. Additionally, some algae are capable of behaving in both autotrophic and heterotrophic modes. These are called mixotrophic algae [104, 112].

Algae range from unicellular to multicellular forms [105]. Some algae are motile while others are nonmotile. Moreover, they may exist as colonies, filaments, or amoeboids [104]. Based on their internal structure, algae cells are generally categorized into eukaryotes and prokaryotes. Prokaryotic cells do not have nuclear membrane-bound DNA, organelles and other membranous structures as eukaryotic cells. As shown in Table1, almost all the algae are eukaryotes. In eukaryotes, microalgae cells consist of cell wall, plasma membrane, cytoplasm, nucleus, and organelles such as mitochondria, lysosomes, and golgi.

**Table 1:** Summary of different algal groups classification for different habitat types [113]

| Kingdom | Division | Habitat | | | |
|---|---|---|---|---|---|
| | | Marine | Freshwater | Terrestrial | Symbiotic |
| Prokaryota | Cyanophyta | Yes | yes | Yes | yes |
| | Prochlorophyta | Yes | yes | Not detected | yes |
| Eukaryota | Glaucophyta | Not detected | yes | Yes | yes |
| | Rhodophyta | yes | yes | Yes | yes |
| | Heterokontophyta | yes | yes | Yes | yes |
| | Haptophyta | yes | yes | Yes | yes |
| | Cryptophyta | yes | yes | Not detected | yes |
| | Chlorarachniophyta | yes | Not detected | Not detected | yes |
| | Dinophyta | yes | yes | Not detected | yes |
| | Euglenophyta | yes | yes | Yes | yes |
| | Chlorophyta | yes | yes | Yes | yes |

As shown in Table 2, microalgae oil contents are usually between 20–50% of dry algae biomass weight. However, many microalgae oil content may exceed 80% of dry algae biomass weight [49, 101, 133, 134]. Besides, microalgae can grow very fast by doubling biomass in 24 hours, and during exponential growth phase they can double their biomass in about 3.5 hours [49, 111, 134, 135].

**Table 2:** Oil content of some common microalgae [49, 114]

| Microalgae | Oil content (% dry biomass weight) |
|---|---|
| Botryococcus braunii | 25–80 |
| Chlorella protothecoides | 23–30 |
| Chlorella vulgaris | 14–40 |
| Crypthecodinium cohnii | 20 |
| Cylindrotheca sp. | 16–37 |
| Dunaliella salina | 14–20 |
| Neochloris oleoabundans | 35–65 |
| Nitzschia sp. | 45–47 |
| Phaeodactylum tricornutum | 20–30 |
| Schizochytrium sp. | 50–77 |
| Spirulina maxima | 4–9 |
| Tetraselmis suecia | 15–23 |

Algal species may change their composition, shape, and color based on growing culture and growth condition such as light, nutrients, temperature, and acidity, pH. It is well known that using stressful environment may cause algae to store more oil.

Unlike glycerolipids that are found in membranes under optimal conditions, many microalgae alter towards accumulations of neutral lipids in form TAG [136]. Microalgae composition is species specific and varies between different microalgae depending on nutrient, salinity, medium pH., temperature, light intensity, and growth phase. In all cells, lipid and fatty acids are constituents that act as membrane compounds, storage product, metabolites, and energy source. It is known that under stress condition, photosynthesis activity decreases; therefore, lipid synthesis occurs. Most of microalgae-produced oils having fatty acid constitutions similar to most common vegetable oils [137].

In general, lipids may include neutral lipids (nonpolar), polar lipids, wax esters, sterols, and hydrocarbons as well phenyl derivatives [138]. Major part of nonpolar lipids of microalgae is TGs and FFA. Typically, algae lipids have a carbon number range $C_{12}$–$C_{22}$. Most of fatty acids found in algae lipids are straight chain with even number of carbon atoms. They may be either saturated or unsaturated [115]. Table 3 gives a summary of the range of lipid reported in different algae species.

**Table 3:** Fatty acid composition of lipids of different microalgae [115]

| Fatty acid | Spirulina platensis | S. maxima | Scenedesmus obliquus | C. vulgaris | Dunaliella bardawil |
|---|---|---|---|---|---|
| $C_{12:0}$ | 0.04 | traces | 0.3 | — | — |
| $C_{14:0}$ | 0.7 | 0.3 | 0.6 | 0.9 | — |
| $C_{14:1}$ | 0.2 | 0.1 | 0.1 | 2 | — |
| $C_{15:0}$ | traces | traces | — | 1.6 | — |
| $C_{16:0}$ | 45.5 | 45.1 | 16.0 | 20.4 | 41.7 |
| $C_{16:1}$ | 9.6 | 6.8 | 8.0 | 5.8 | 7.3 |
| $C_{16:2}$ | 1.2 | traces | 1.0 | 1.7 | — |
| $C_{16:4}$ | — | — | 26.0 | — | 3.7 |
| $C_{17:0}$ | 0.3 | 0.2 | — | 2.5 | — |
| $C_{18:0}$ | 1.3 | 1.4 | 0.3 | 15.3 | 2.9 |
| $C_{18:1}$ | 3.8 | 1.9 | 8.0 | 6.6 | 8.8 |

| | | | | | |
|---|---|---|---|---|---|
| $C_{18:2}$ | 14.5 | 14.6 | 6.0 | 1.5 | 15.1 |
| $-C_{18:3}$ | 0.3 | 0.3 | 28.0 | — | 20.5 |
| $-C_{18:3}$ | 21.1 | 20.3 | — | — | — |
| $C_{20:2}$ | — | — | — | 1.5 | — |
| $C_{20:3}$ | 0.4 | 0.8 | — | 20.8 | — |
| Others | — | — | 2.5 | 19.6 | — |

Microalgae are classified into four main classes according to their pigment components: diatoms (Bacillariophyceae), green algae (Chlorophyceae), blue-green algae (Cyanophyceae), and golden algae (Chrysophyceae) [102, 103, 108, and 139]. Table 4 gives a brief description of each division.

**Table 4:** Summary of different microalgae divisions [104, 113, and 116]

| Division | Examples | Occurrence | Photosynthesis pigments | Reproduction |
|---|---|---|---|---|
| Diatoms | Coscinodiscus granii | | | |
| | Tabellaria | | | |
| | Amphipleura | (i) Oceans | (i) Chlorophylls a and c (ii) B-carotene | (i) Vegetative (binary fission or fragmentation) |
| | Thalassiosira baltica | (ii) Freshwater | | (ii) Asexual (akinete, exospores, endospores or homospores)(iii) Sexual (isogamous, anisogamous or oogamous) |
| | Skeletonoma | (iii) Brackish water | | |
| | Chaetoceros | | | |
| | Cyclotella | | | |
| | Chlorella sp. | | | |
| Green algae | C. vulgaris | | | |
| | C. protothecoides | | | |
| | | (i) Oceans | | |
| | S. obliquus | (ii) Freshwaters | () Chlorophylls a and b | (i) Vegetative (binary fission or fragmentation) |
| | Haematococcus pluvialis | (iii) Moist | | (ii) Asexual (akinete or exospores or endospores or homospores) |
| | Nannochloris | (iv) Terrestrial habitats. | | (iii) Sexual (isogamous, anisogamous or oogamous) |
| | D. salina | | | |
| | B. braunii | | | |

| | | | (i) Divided in to two groups | |
|---|---|---|---|---|
| Blue-green algae | S. platensis | (i) Freshwater | (ii) Most species have chlorophyll a as only form of chlorophyll and phycobilins as pigments | (i) Vegetative (binary fission and fragmentation) |
| | Synechococcus | (ii) Marine | | (ii) Asexual (akinete or exospores or endospores and homospores) |
| | Cyanidium | (iii) Terrestrial | | |
| | Oscillatoria | (iv) Symbiotic | | |
| | Anabaena cylindrical | (v) Associations | (iii) Some have two forms of chlorophyll aand b and lack phycobilins | |
| Golden algae | Isochrysis galbana | (i) Fresh water | (i) Chlorophylls a and b | (i) Asexual (zoospores or aplanospore, hypnospores) |
| | Dinobryon balticum | (ii) Marine | (ii) Some have chlorophylls e carotene aand c | (ii) Sexual (isogamous or anisogamous or oogamous) |
| | Uroglena americana | (iii) Terrestrial | | |

Microalgae biomass contains three main components: protein, carbohydrates, and lipids and, therefore, can be used in different applications ranging from food products to biofuels. They are usually used as animal feed [140], human health food [104, 141], and as biofertilizer [49]. Additionally, microalgae can be used for atmospheric $CO_2$ mitigation. It was reported that there are over 40,000 species of algae [136], but only limited number of these have been studied and have commercial significance [142].

# Potential of Using Microalgae as Feedstock for Biodiesel Production

Biodiesel production from microalgae oil is more promising and sustainable alternative to previously mentioned feedstocks (Section 2.2) [143]. Compared to plants, algae do not compete with food crops and have higher energy yields per area than terrestrial crops. They present a good source of renewable biofuels, which include methane via anaerobic digestion of algal biomass, biodiesel derived from microalgae oil, and biohydrogen via photobioproduction [133, 144, 145]. The focus of this paper is on biodiesel.

Using microalgae as a fuel source is not new; it was suggested more than 60 years ago by Meier for methane production [136]. However, only recently microalgae received noticeable attention due to increasing environmental concerns. Main advantage of using microalgae as feedstock is their rapid growth potential with short biomass doubling time (3.5 hours) during exponential growth and oil content ranging from 20 to 50% dry weight of biomass for numerous microalgae species, as shown in Table 2. Other major advantages and features of using microalgae are the following.

- Ability to grow in nonarable land [49, 126, 144, 146, 147] where they can be cultivated on lands that is unsuitable for agriculture, that is, waste land [103, 148]. Therefore, biodiesel production would not be in any way competing with food production [106, 149].

- Can be cultivated in saline and brackish environments leading to reduction in fresh water load [111].

- (Daily harvesting [49, 126, 150] and short harvesting cycle [106, 111] in comparison to crop plants.

- High photosynthetic efficiency due to their simple structure [111, 147, 148].

- Reduction in major greenhouse gas contributor, by utilizing $CO_2$ from industrial flue gasses [49,126, 146, 149–151]. Thereby, they are considered as $CO_2$ fixers [152].

Table 5 compares different sources of biodiesel and their oil yield per area. Chisti [49] mentioned that in order to satisfy with the US demand for transportation fuel, the biodiesel industry would need to produce 530 billion liters annually. As can be found from Table 5, the most feasible biodiesel source for the US is microalgae.

**Table 5:** Comparison between different biodiesel sources [49]

| Crop | Oil yield (L/ha) | Land area needed (M ha)[a] | Percent of existing US cropping area[a] |
|------|------------------|----------------------------|------------------------------------------|
| Corn | 172 | 1540 | 846 |
| Soybean | 446 | 594 | 326 |
| Canola | 1190 | 223 | 122 |
| Jatropha | 1892 | 140 | 77 |
| Coconut | 2689 | 99 | 54 |
| Oil palm | 5950 | 45 | 24 |
| Microalgae[b] | 136,900 | 2 | 1.1 |
| Microalgae[c] | 58,700 | 4.5 | 2.2 |

[a]For meeting 50% of all transport fuel needs of the United States

[b]70% oil by wt in biomass

[c]30% oil by wt in biomass.

Consequently, biodiesel production from microalgae is considered to be the best efficient feedstock for biodiesel production to displace conventional feedstock's and meet global demand of fuel [153].

To use microalgae for the production of biodiesel, several processes have to be carried out. These consist of strain selection, cultivation, harvesting, extraction of the oil, and production of biodiesel from extracted oil, in which each step can be accomplished with various technologies. These steps are detailed in the following sections.

# Microalgae Oil Production Systems

## *Microalgal Strain Selection*

Microalgae come in a variety of strains; each has different proportions of lipid, protein, and carbohydrates contents. From over 3,000 collected, screened, and characterized algal strains in the National Renewable Energy laboratory (NREL) sponsored project [50], selection of the most suitable strain needs certain parameters evaluation. These parameters include oil content, growth rate and productivity, strain adaptableness, and withstanding to different weather conditions such as temperature, salinity and pH, and high $CO_2$ sinking capacity and provide valuable coproducts [108]. Thus, right strain selection is critical.

Numerous researches have been carried out on different species tolerance. Many of them were found to be suitable for biodiesel production. As shown in Table 6, some microalgae have high lipids content such asNannochloropsis sp., and others have high protein contents like C. protothecoides (autotrophic) while others have high carbohydrates content like Oscillatoria limnetica under normal conditions. Among possible microalgae strains for biodiesel productions are Chlorella species. As can be seen from Table 2, C. protothecoides oil content is roughly 15% (% dry weight) under control conditions, but this can reach around 44% [154, 155], 53% [156], and 55% [120] when grown heterotrophically.

**Table 6:** Chemical composition of various microalgae (% dry weight)

| Microalgae | Carbohydrates | Protein | Lipids | Reference (s) |
|---|---|---|---|---|
| Chaetoceros muelleri | 19.3 | 46.9 | 33.2 | [117] |
| I. galbana | 26.8 | 47.9 | 14.5 | [118] |
| Chaetoceros calcitrans | 27.4 | 36.4 | 15.5 | [118] |
| Isochrysis sp. | 12.9 | 50.8 | 20.7 | [119] |
| Prymnesiophyte (NT19) | 8.4 | 41.3 | 14.7 | [119] |
| Rhodomonas (NT15) | 6.0 | 57.2 | 12 | [119] |
| Cryptomonas (CRF101) | 4.4 | 44.2 | 21.4 | [119] |
| Chaetoceros (CS256) | 13.1 | 57.3 | 16.8 | [119] |
| C. protothecoides [a] | 10.6 | 52.6 | 14.6 | [120–122] |
| C. protothecoides [b] | 15.4 | 10.3 | 55.2 | [120, 121] |
| Microcystis aeruginosa | 11.6 | 30.8 | 12.5 | [122] |
| Nannochloropsis sp | 29.0 | 10.7 | 60.7 | [123] |
| S. obliquus | 15 | 50.0 | 9.0 | [124] |
| Oscillatoria limnetica | 50 | 44.0 | 5.0 | [124] |
| B. braunii | 18.9 | 17.8 | 61.4 | [125] |
| Botryococcus protuberans | 16.8 | 14.2 | 52.2 | [125] |

[a]Autotrophic cultivation

[b]Heterotrophic cultivation.

On the other hand, generally, lower oil strains grow faster than high oil strains. This is due to slow reproduction rate as a result of storing energy as oil not as carbohydrates [157]. In addition, it should be taken into consideration that some microalgae contain high levels of unsaturated fatty acids, which reduce the oxidative stability of the biodiesel produced [158–160].

Rodolfi et al. [126] have screened variety of microalgal strains by evaluating biomass productivity and lipid content in 250-mL flask laboratory cultures (Table 7). Strains that have shown some promise lipid productivity can be further improved genetically. Genetic engineering can improve all aspects of algal production, harvesting, and processing for enhanced biodiesel capabilities.

**Table 7:** Biomass productivity, lipid content, and lipid productivity of 30 microalgal strains cultivated in 250-mL flasks [126]

| Algal Group | Microalgae strain | Habitat | Biomass productivity (g l⁻¹ d⁻¹) | Lipid content (%) | Lipid productivity (mg l⁻¹ d⁻¹) |
|---|---|---|---|---|---|
| Diatoms | Chaetoceros muelleri F&M-M43 | Marine | 0.07 | 33.6 | 21.8 |
| | C. calcitrans CS 178 | Marine | 0.04 | 39.8 | 17.6 |
| | P. tricornutum F&M-M 40 | Marine | 0.24 | 18.7 | 44.8 |
| | Skeletonoma costatum CS 181 | Marine | 0.08 | 21.0 | 17.4 |
| | Skeletonoma sp. CS 252 | Marine | 0.09 | 31.8 | 27.3 |
| | Thalassiosira pseudonana CS 173 | Marine | 0.08 | 20.6 | 17.4 |
| | Chlorella sp. F&M-M48 | Freshwater | 0.23 | 18.7 | 42.1 |
| | Chlorella sorokiniana IAM-212 | Freshwater | 0.23 | 19.3 | 44.7 |
| | C. vulgaris CCAP 211/11b | Freshwater | 0.17 | 19.2 | 32.6 |
| | C. vulgaris F&M-M49 | Freshwater | 0.20 | 18.4 | 36.9 |

| Green algae | Chlorococcum sp. UMACC 112 | Freshwater | 0.28 | 19.3 | 53.7 |
|---|---|---|---|---|---|
| | Scenedesmus quadricauda | Freshwater | 0.19 | 18.4 | 35.1 |
| | Scenedesmus F&M-M19 | Freshwater | 0.21 | 19.6 | 40.8 |
| | Scenedesmus sp. DM | Freshwater | 0.26 | 21.1 | 53.9 |
| | Tetraselmis. suecica F&M-M33 | Marine | 0.32 | 8.5 | 27.0 |
| | Tetraselmis sp. F&M-M34 | Marine | 0.30 | 14.7 | 43.4 |
| | T. suecica F&M-M35 | Marine | 0.28 | 12.9 | 36.4 |
| | Ellipsoidion sp. F&M-M31 | Marine | 0.17 | 27.4 | 47.3 |
| | Monodus subterraneusUTEX 151 | Freshwater | 0.19 | 16.1 | 30.4 |
| | Nannochloropsis sp. CS 246 | Marine | 0.17 | 29.2 | 49.7 |
| Eustigmato-phytes | Nannochloropsis sp. F&M-M26 | Marine | 0.21 | 29.6 | 61.0 |
| | Nannochloropsis sp. F&M-M27 | Marine | 0.20 | 24.4 | 48.2 |
| | Nannochloropsis sp. F&M-M24 | Marine | 0.18 | 30.9 | 54.8 |
| | Nannochloropsis sp. F&M-M29 | Marine | 0.17 | 21.6 | 37.6 |
| | Nannochloropsis sp. F&M-M28 | Marine | 0.17 | 35.7 | 60.9 |
| | Isochrysis sp. (T-ISO) CS 177 | Marine | 0.17 | 22.4 | 37.7 |
| | Isochrysis sp. F&M-M37 | Marine | 0.14 | 27.4 | 37.8 |
| Prymnesio-phytes | Pavlova salina CS 49 | Marine | 0.16 | 30.9 | 49.4 |
| | Pavlova lutheri CS 182 | Marine | 0.14 | 35.5 | 50.2 |
| Red algae | Porphyridium cruentum | Marine | 0.37 | 9.5 | 34.8 |

# Biomass Production

The main way to produce microalgal biomass is the cultivation. For commercial biomass production, microalgal biomass must be easily cultivated in the required scale. Microalgae cultivation can be carried out either via photoautotrophic methods in open systems (open-ponds) [153, 161] or closed systems (photobioreactors) [102, 161, 162], or via heterotrophic methods [120, 154]. All methods have their advantages and disadvantages; therefore, investigators disagree about which of the methods and systems is more favorable. Choosing best biomass production method or system depends on the selected algal strain and its integration with appropriate downstream processing which is the means for affordability and scalability of biodiesel production.

## Photoautotrophic

As mentioned previously, photoautotrophic microalgae need light and carbon dioxide as energy and carbon sources, respectively. Thus, photoautotrophic algae cultivation is carried out in the presence of light in open ponds and photobioreactors. Open ponds are the most commonly used systems, and their structure has been well documented. Open ponds are made of a closed loop with recirculation channels. A paddlewheel that continuously operates is usually used to prevent sedimentation and provide mixing. During daylight, the culture is fed continuously in front of the paddlewheel where the flow begins and circulates through the loop to the harvesting point. On completion of the circulation loop, broth is harvested behind the paddlewheel [49, 102, 161, 163]. Inclined, circular, and raceway ponds are operated at large scale. On the other hand, photobioreactors are closed bioreactors, which are designed as tubular, plate, or bubble column reactors. Among these, the most common type is tubular photobioreactors. These consist of less than 0.1 m diameter transparent tubes made from plastic or glass. Tube diameter is a critical design criteria as light does not penetrate too deeply in dense culture broths [49]. This leads to $O_2$ accumulation and thus inhibits the photosynthesis process. Typically, open ponds are the preferred large scale cultivation system [49]. This is due to their simplicity and low construction and capital costs [163]. However, these systems are open to the atmosphere, which lead to water evaporation and unwanted species contaminations. Besides, cell's poor utilization of light and low mass productivity, due to the low $CO_2$ deficiencies and inefficient mixing, are other limitations [151, 164]. Therefore, for water, energy, and chemicals saving purposes, photobioreactors have been proposed, but they are not yet commercialized. Main advantages of using photobioreactors are better algal culture and environment controlling [163], large surface to volume ratio, less water evaporation, better isolation from outside contaminations, and higher mass productivity [158]. However, photobioreactors are usually made of plastic, and UV deterioration of plastic surface is the main disadvantage. In addition, biofilm formation will require periodic cleaning [165]. Table 8 shows a comparison between the two photoautotrophic cultivation methods. For a cost-effective cultivation, a combination of the two previous mentioned systems, referred to as hybrid system, is the most logical choice [153]. In this type of systems, microalgal strain with high oil content is grown in photobioreactors in nutrient and $CO_2$-rich conditions firstly to promote rapid reproduction; then the microalgae enter an open system with limited nutrient to encourage oil production [166]. This process has been successfully verified by Huntley and Redalje [167]. In addition to microalgae strains influence on oil accumulation, cultivation parameters like temperature, light intensity, pH, water salinity, and nitrogen sources also influence oil production. It has been reported that the lipid content in various microalgae strains from Chlorella species increased when growing in low-nitrogen media compared to nitrogen-rich media [168, 169]. However, in these low-nitrogen media, a reduction in growth rate was reported. Similar results were also found by Widjaja et al. [170]. Since the cell needs sufficient nitrogen for growth, the cell production and division may reduce in the low-nitrogen media. However, carbon metabolism continues leading to utilize more energy for oil production rather than biomass growth [50, 171]. Other factors like $CO_2$, light intensity, and temperature also significantly affect microalgae lipid content and composition. Renaud et al. [119] investigated the effect of temperature within the range of 25 to 35°C onRhodomonas sp., Chaetoceros sp., Cryptomonas sp., and Isochrysis sp. growth rate and lipid content. Their results showed that optimum growth temperature was 25–27°C for Rhodomonas sp., and 27–30°C forCryptomonas sp., Chaetoceros sp., and Isochrysis sp. Only Chaetoceros sp. was able to grow at 33 and 35°C. With the intent of providing sufficient light to the cultivation systems, open ponds are usually made shallow, and tabular reactors are made small in diameters. Tang et al. [172] studied the influence of the above mentioned parameters on Dunaliella tertiolecta growth, lipid content, and fatty acid composition. It was reported that increasing light intensity increases cell growth rate regardless of the light source. On the other hand, as for the $CO_2$ effect, the highest growth rate was found when $CO_2$ concentration was in the range of 2 to 6%.

**Table 8:** Comparison between open ponds and photobioreactors

| Method | Advantages | Limitations |
|---|---|---|
| Open ponds | (i) Simple (ii) Cheap (iii) Easy to operate and maintain (iv) Low capital cost | (i) Poor light utilization (ii) Difficulties in light and temperature controlling (iii) Water evaporation (iv) Foreign species contaminations (v) Lower mass productivity |
| Photobioreactors | (i) High surface to volume ratio (ii) Higher mass productivity (iii) Less contaminations (iv) Less water losses (v) Better light utilization | (i) Scalability problem (ii) Costly (iii) Complex (iv) Cells damage cases (v) Biofilm formation |

Heterotrophic CultivationUnlike photoautotrophic microalgae, heterotrophic species are cultivated in a dark environment by utilizing organic carbon as carbon and energy sources [109, 173]. Heterotrophic cultivation method depends on the microalgae ability to eliminate light requirement and assimilate organic carbon [174]. This solve light limitation problem that appears with photoautotrophic cultivation methods. However, not all microalgae are able to assimilate the organic carbon. Thus, this cultivation method has been studied in a limited number of microalgae species [120, 154, 175]. It has been reported that heterotrophic cultivation provides high oil content and high biomass productivity [120, 121, 154, 176, 177]. Liu et al. [178] compared lipid content of Chlorella zofingiensis cultivated under heterotrophic and photoautotrophic conditions. Lipid content of 51 wt% and 26 wt% were obtained, respectively. Liang et al. [174] compared C. vulgaris cell growth rate and lipid productivity under autotrophic and heterotrophic conditions, evaluated glucose, acetate, and glycerol carbon sources uptakes. Table 9 illustrates the results obtained.

**Table 9:** Lipid content, biomass, and lipid productivities of C. vulgaris grown autotrophically and heterotrophically on different carbon sources

| | Heterotrophic cultivation | | | Autotrophic cultivation |
|---|---|---|---|---|
| | Acetate | Glucose | Glycerol | |
| Biomass productivity (mg l$^{-1}$ d$^{-1}$) | 87 | 151 | 102 | 10 |
| Lipid content (%) | 31 | 23 | 22 | 38 |
| Lipid productivity (mg L$^{-1}$ d$^{-1}$) | 27 | 35 | 22 | 4 |

## *Harvesting Technologies*

After algal cultivation, biomass needs to be separated from the culture medium using one or more solid-liquid separation steps. Due to the microalgae small size (3–30 µm) [102, 179, 180] and cultures medium dilution (less than 1 g L$^{-1}$), microalgae need to be concentrated to simplify the lipid extraction step. Biomass recovery is difficult [181] and require dewatering using suitable harvesting method [49, 151, 182]

Usually, microalgae are harvested by centrifugation, filtration, or sedimentation. Sometimes these require a pretreatment, flocculation step to improve recovery efficiency [183, 184]. Table 10 summarizes the advantages and disadvantages of each method.

**Table 10:** Advantages and disadvantages of different microalgal harvesting methods

| Method | Advantage | Disadvantage |
|---|---|---|
| Flocculation | High recovery yield (up to 22 TTS) | Flocculants may be expensive |
| | Low energy requirement | Contamination issue may occur |
| | | Marine environment high salinity may inhibit the process |
| | | Long process period |

| Centrifugation | Reliable | Energy intensive |
|---|---|---|
| | Corrosion resistance | Expensive |
| | Easy cleaning | High speed may deteriorate the cell |
| | Rapid | Cannot be used for species <30 μm |
| Filtration | Reliable | Filters may need to be replaced periodically |
| | Able to collect species of low density | Membrane blockage |
| | | High maintenance cost |
| | | May be slow |
| | | Head loss |

To conserve energy and reduce costs, algae are often harvested in a two-step process. In the first, algae are concentrated by flocculation where diluted culture is concentrated to about 2–7% total suspended solids [108, 183]. In the second step, cells are further concentrated using conventional harvesting methods to get an algae paste of 15–25% total suspended solids [183]. Algae harvesting cost can be high due to their low mass fraction and algal cell negative charge [182]. It is reported that microalgal cell recovery accounts for at least 20–30% of total biomass production cost [184, 185]. Harvesting technique selection depends on microalgal cells size and density, biomass concentration, culture conditions, and value of target product [102, 186].

## Flocculation

Flocculation is a process that collects dispersed cells into aggregate to form large particles that facilitate cell broth separation by addition of chemical additives (flocculants). It is considered as pretreatment stage preceding the main harvesting process [184]. The main problems facing the flocculation step are the high cost and toxicity of the flocculent [187]. To endorse flocculation, chemical additives that bind algae or affect interaction between algae have to be used. There are two main types of flocculants inorganic and organic polymer. A large number of chemicals have been tested as flocculants for microalgal flocculation where the most effective one was aluminum sulfate and certain cationic polymers [187]. Numerous reports have been published concerning the flocculation of algal biomass. Among them is the work of Tenney et al. [188] which looked into fresh water microalgae flocculation using organic polyelectrolytes where extent of microalgal flocculation was determined. Cationic polyelectrolyte polyamine was found to flocculate the algae successfully at an optimum dose of 2.5 mg/L.

## Filtration

Filtration separation method makes use of a permeable medium that has an ability to retain the biomass and allows the liquid to pass through. Surface and depth filtration systems are the two known types of filtration. In surface filtration, solids are deposited on the filter medium whereas in the depth type solids are deposited within the filter medium [189]. This is satisfactory for recovering large microalgae and not for algae that size approach bacterial dimension.

## Sedimentation

Sedimentation is a technique that separates microalgae biomass suspension into a concentrated slurry and clear liquid based on gravity action and particle diameter. If the biomass to be separated is small in size, settling rate will be low, and flocculants addition will be inevitable. This is a low cost process; however, its reliability is low.

## Centrifugation

Almost all types of microalgae can be separated from the culture medium by centrifugation. A centrifuge is mainly a sedimentation tank with an enhanced gravitational force, by centrifugation, that increases the rate of sedimentation. Biomass recovery depends on biomass residence time within centrifugal field, settling rate, and distance [184]. Centrifugal recovery can be rapid, but it is energy intensive. Nevertheless, this process is a preferred method of recovering algal cells [183, 184]. Currently, there is no low cost harvesting method for all strains. Table 10 summarizes advantages and disadvantages of each harvesting method.

## Drying

Following harvesting of the algal biomass, algal slurry moisture content has to be reduced to at least 10% by drying and dehydration. Numerous methods have been employed for drying. Most common methods are sun drying, spray drying, drum drying, and freeze drying. Again, best drying method selection depends on required operation scale and desired product value.

In biodiesel production, lipid-rich feedstock with low water content is required; therefore, microalgae drying has to be carried out. However, drying step is energy intensive, which adds to the cost complexity of the overall production process. Various drying systems differ in both energy and cost requirements.

Sun drying is an old and cheap drying method that can be performed easily by exposure to a solar radiation source. However, it takes long drying time, requires large drying surface area, and might result in loss of products. In addition, sun drying, unlike drum drying, does not have any sterilization effect of the dried sample. On the other hand, spray drying is a method that can be used for high value products, but it has the disadvantage of being expensive and might cause significant deterioration of algae [184]. In contrast, freeze drying has been commonly used by many investigators. Freeze drying has the advantage of breaking up species cells and turning them into fine powder that makes homogenization unnecessary [190]. Belarbi et al. [191] reported that freeze-dried sample can be subjected easily for oil extraction without cell disruption. Freeze driers have been used in algae lipid extraction to extract lipid from I. galbana [192], P. tricornutum[193], C. vulgaris [194], S. platensis [195], and Chlorella sp. [196]. Table 11 summarizes advantages and disadvantages of each technique. However, freeze drying is a slow process and requires very high capital investment.

**Table 11:** Comparison between common four microalgae drying methods

| Method | Advantages | Disadvantage |
|---|---|---|
| Sun drying | (i) Cheap (no running cost, low capital cost) | (i) Difficult (ii) Slow (iii) Weather dependent (iv) Require large surface (v) Contamination |
| Spray drying | (i) Fast (ii) Continuous (iii) Efficient | (i) Cost intensive (ii) Species deterioration (i.e. pigments) |

| Drum drying | (i) Fast (ii) Efficient (iii) Sterilization advantage | (i) Cost intensive |
|---|---|---|
| Freeze drying | (i) Gentle | (i) Slow process (ii) Cost intensive |

## Oil Extraction

As stated previously, effective extraction requires concentrated algae solution. Thus, a high degree of algae concentration which takes place in the harvesting step is necessary before biomass lipid extraction. Typically, there are four well-known methods to extract oil from microalgae: (1) expeller/press, (2) solvent extraction with hexane, (3) subcritical water extraction, and (4) supercritical fluid extraction.

The recovering of intracellular products like oils from microalgae is usually difficult due to the cell wall robust structure [151, 197]. Therefore, prior to lipid extraction, algae cell has to be disrupted to a degree that facilitates extraction step [184, 198]. Several methods can be used to disrupt cell membrane. They include homogenizer, bed mill, ultrasound, autoclaving, freezing, and osmotic shock [151]. Among them, homogenizers and bed mills are often preferred because of their short residence time and lower operating costs [175]. Chisti and Moo-young [197] reviewed microbial cell disruption for intracellular products.

## Mechanical Extraction

Mechanical oil extraction includes expeller press and ultrasonic extraction. In pressing technique, the presser crushes and pushes the oil out of the dry microalgal biomass. Despite expeller simplicity and lower investment cost, low oil recovery yield, high power consumption, and maintenance cost are the limitations [199]. In ultrasonic technique, shock waves break the walls and release oil to solvent. These waves are created when bubbles (created by ultrasonic wave associated from ultrasonic reactor) collapse near cell wall. Ultrasonic extraction has an advantage of being fast and efficient; at the same time it needs large amount of solvent, especially in case of low sample concentration. That's because at low concentrations, samples need to be extracted more than once using new fresh solvent [200].

## Chemical Extraction

The well-known concept of "like dissolve like" is the basic of the Bligh and Dyer [201] solvent extraction method. This method is the most widely used for extracting lipids from microalgae, wherein hexane is one of the most widely used solvent due to its high extraction capability and low cost. For successful lipid extraction using an organic solvent, the solvent must be able to penetrate through the matrix to contact and dissolve the lipid. When hexane is used as a solvent, it is mixed with the algal biomass and is then separated by filtration. The solvent has to be separated from the extracted oil using distillation which is energy intensive. Miao and Wu [120] reported that large amount of microalgal oil was efficiently extracted from C. protothecoides using n-hexane. Beside, cosolvent combinations have been used by many other investigators [192, 193, 202, 203]. Hexane/ethanol and hexane/isopropanol cosolvents have been commonly used in microalgal lipid extraction. The polar solvent, which is the alcohol, is first added to disrupt the algal cell membrane. This will enhance the ability of the hexane to extract almost all the lipids. The cosolvent is then removed by liquid-liquid extraction with water. The hexane solvent extraction method can also be used in combination with the oil press/expeller mechanical method. After extracting the oil from the algae using the expeller, the remaining pulp is then mixed with hexane in order to remove any remaining oil. In this combined method, more than 95% of the total oil present in the algae is extracted [204]. The selection of lipid extraction methods depends on the extraction efficiency. Therefore, a method of high performance, such as chemical extraction, is favored over the less efficient methods, such as mechanical extraction, despite the organic solvents negative environmental impacts. To avoid the environmental impacts of using organic solvent, nontoxic solvents have been suggested, such as subcritical water (SCW) and SC-$CO_2$. The SCW extraction operates at temperatures just below the critical temperature, 374°C, and at high pressure, usually from 10 to 60 bar, that maintains the water in liquid form. At these conditions, water becomes less polar, and lipids can be solubilized easily. Additionally, using water at subcritical condition can eliminate the dewatering step, and high-quality product within short extraction time can be achieved [205]. However, reaching the above mentioned temperature requires large energy consumption. On the other hand, supercritical fluids extraction makes use of fluid's salvation power enhancement when reached above their critical point. Due to supercritical carbon dioxide's preferred critical properties, low toxicity, biodegradability, and availability, it has been used to extract many desired compounds from solid matrix. Other attractive point of using SC-$CO_2$ as extraction solvent is that after extraction, solvent and product can easily be separated once the temperature and pressure are lowered to atmospheric conditions.

# Microalgae Oil Production Costs

The idea of producing biodiesel from microalgae was the main focus of NREL project [50]. It is not much different from other biodiesel produced from vegetable oils, animal fats, or waste cooking oils. It was reported that biodiesel from vegetable oil and waste grease roughly costs $ 0.54 to $ 0.62/L and $ 0.34 to $ 0.42/L, respectively [206]. Chisti [49] reported that biodiesel from palm oil almost costs $ 0.66/L and in year 2006 petrodiesel average price was $ 0.49/L, which added about $ 0.14 to palm oil cost and 35% more than petrodiesel price. The objective is therefore to reduce the product cost to at least $ 0.48/L ignoring the effect of tax on biodiesel. However, the estimated cost of biodiesel increases to $ 0.72–1.4/L using microalgae with 70 wt% and 30 wt% (per dry-weight) oil content, respectively [49].

The high cost of biodiesel comes mainly from the high feedstock cost; 60–90% of biodiesel cost is estimated to be from the cost of the feedstock [7]. Therefore, looking for alternatives that are cheap became essential. From that point, microalgal oil production should be enhanced. Microalgae growth requires light, $CO_2$, water, and salt utilization. To minimize production cost, oil production must rely on maximum available mentioned requirements. Therefore, using water from waste water treatment units that contain required growth nutrients and salts and diverting $CO_2$ from power plants is desirable and beneficial. Other attributers are development of low harvesting process by genetic engineering, improvements in photobioreactors design, and coproduction of other high values products from residual biomass after lipid extraction [184].

# SUPERCRITICAL FLUIDS

Supercritical fluids (SCFs) are fluids at pressures and temperatures above their critical values. Critical values represent the highest temperature and pressure at which the substance exists as a vapor and liquid in equilibrium. This can be simply clarified from supercritical fluids phase diagram (Figure 2).

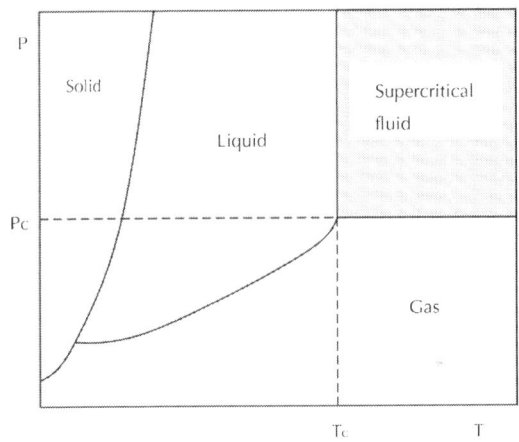

**Figure 2:** Pure component phase diagram.

As shown in Figure 2, there are three single phases, solid, liquid, and gas, where a substance may occur. If a mixture of two, or more, phases exists in these regions, a separation between the phases is distinct, as a result of the difference in properties of the different phases. In Figure 2, the solid curves between phases indicate the coexistence of two phases. On the other hand, at a point beyond the critical point neither fusion, as a result of pressure increase, nor vaporization, as a result of temperature increase, will take place, which was defined earlier as a supercritical region. Table 12 presents the values of the critical temperatures and pressures of selected fluids most commonly used as extraction solvents [127].

**Table 12:** Critical properties of common solvents [127]

| Fluid | Critical temperature (°C) | Critical pressure (bar) |
|---|---|---|
| Xenon | 16.7 | 59.2 |
| Carbon dioxide | 31.1 | 72.8 |
| Ethane | 32.4 | 49.5 |
| Nitrous oxide | 36.6 | 73.4 |
| Chlorodifluoromethane | 96.3 | 50.3 |
| Ammonia | 132.4 | 115.0 |
| Methanol | 240.1 | 82.0 |
| Water | 374.4 | 224.1 |

In the SCF state, a solvent displays properties which are intermediate to those of liquid and gaseous states; SCFs have more desirable transport properties than liquids and better solvent properties than gases. The liquid-like density of a SCF gives high solvation power and facilitates solubility while the gas-like diffusivity gives excellent transport properties, which increases the rates of transfer from the substrate matrix to the SCF solvent as compared to that of liquid organic solvents [207]. Moreover, the low viscosity of SCFs which is close to that of the gases is an additional advantage. This last property gives rapid solvent penetration into a solid matrix [208].

# Supercritical Carbon Dioxide as a Candidate Solvent

The ability of supercritical carbon dioxide, SC-$CO_2$, to extract a solute depends on the compounds functional groups, molecular weight, and polarity. Near to its critical point, $CO_2$ is a good solvent for nonpolar to slightly polar solutes with low molecular weight. It is an inert at most conditions, inexpensive, nontoxic, and environmentally friendly [43, 209]. Moreover, when using SC-$CO_2$ as a solvent, no solvent residue remains in the extract since $CO_2$ is in a gas phase at the ambient conditions. The critical temperature and critical pressure of $CO_2$ are 31.1°C and 72.8 bar, respectively, which are not extremely high. SC-$CO_2$ has been identified as a good alternative solvent for a number of applications including separation and reaction.

## SC-CO₂: *Extraction Solvent*

Extraction is the process of removal of a solute from a matrix using a solvent which is able to dissolve the desired solute. This involves contacting the matrix with the solvent either in a single stage or in multiple stages for certain period of time and then separating the solvent. During extraction period, the solute transfers from the matrix to the solvent. Required time to achieve successful extraction depends on the solubility. That depends on extraction temperature, contact area between the solute and solvent, solvent viscosity, and solvent flow rate.

Other conventional solvent extraction techniques suffer from several drawbacks such as long extraction time and high solvent consumption, in addition to being labor intensive, difficult to automate, and often require a postextraction cleanup [210]. With these drawbacks, supercritical fluid extraction, SFE, has been proposed using the extraction solvent in its supercritical state.

SCFs were first observed more than a century ago in 1822. However, it has been developed as a novel separation technique only in the past two decades. From an economical point of view and in order not to thermally alter the properties of the extracted materials, SCFs are mostly used as in the approximate range of temperature up to 1.2 times the critical temperature, , and pressure up to 3.5 times the critical pressure [211].

Although a number of substances could serve as solvents, $CO_2$ is the most common. SC-$CO_2$ has many applications, especially in food processing, which include decaffeination of coffee and tea, production of hops extracts, flavors extract from herbs, and extraction of edible oils. Friedrich and Pryde [212] extracted oil from soybeans using SC-$CO_2$ and achieved a yield almost to that using n-hexane. In order to extract polar compounds from a matrix, polar supercritical fluid should be used. Thus, using a nonpolar solvent, $CO_2$ sometimes faces difficulties to extract certain compounds from a sample matrix. To overcome this limitation, modifier fluids can be used to increase extraction efficiency. Among all modifiers tested, methanol was the most commonly used by investigators such as Tonthubthimthong et al. [213] who extracted nimbin from neem seeds. Brewer et al. [214] who extracted cocaine from human hair and Aghel et al. [215] who extracted pennyroyal essential oil using SC-$CO_2$.

Due to the attractive features of SC-CO$_2$, it has been used and assessed to extract lipids from different strains of microalgae [128–132, 216]. Maximum yields of 13, 9, 25, 8, 6, and 3% have been reported from C. vulgaris[128, 132], C. cohnii [216], Nannochloropsis sp. [129], S. platensis [130], chlorococum sp. [131], and S. maxima [132], respectively. The lipids were extracted from dried biomass in a temperature range of 40°C–80°C and pressure range of 100–550 bars. The lower extract yield was due to the low lipid content of grown biomass.

The extraction efficiency of SC-CO$_2$ was compared to conventional solvent extraction methods. Table 13shows the extraction yields of lipids, defined as amount of extracted lipids per dry biomass weight, extracted from different strains of microalgae using SC-CO$_2$, as compared to that of conventional solvent extraction. As shown, similar yields were reported when using SC-CO$_2$ and n-hexane for extracting lipids from S. platensis [130] and S. maxima [132]. However, 25–40% lower yields were reported when comparing SC-CO$_2$to n-hexane and acetone extractions, from C. vulgaris [128]. A lower yield was also reported when comparing SC-CO$_2$ and ethanol extractions from S. maxima [132]. However, other studies showed a better performance than n-hexane from chlorococum sp. and Nannochloropsis sp. [129, 131].

**Table 13:** Comparison of SC-CO$_2$ performance and other conventional extraction solvents on lipids extraction yields from microalgae biomass

| Microalgae species | SC-CO$_2$ | Other conventional solvents | | | Reference |
|---|---|---|---|---|---|
| | | Acetone | Ethanol | n-Hexane | |
| C. vulgaris | 13.3 | 16.8 | — | 18.5 | [128] |
| Nannochloropsis sp | 25 | — | — | 23 | [129] |
| S. platensis | 7.8 | — | — | 7.7 | [130] |
| chlorococum sp | 5.8 | — | — | 3.2 | [131] |
| S. maxima | 2.5 | 4.7 | 5.7 | 2.6 | [132] |

To further enhance the SC-CO$_2$ extraction yields, the use of a cosolvent has been suggested. SC-CO$_2$ with 10% ethanol as a cosolvent for lipid extraction from S. maxima has been reported [217]. By doing so, the extraction yield increased by 24% from 32% to reach 40%. This enhancement was explained by ethanol destruct effect on microalgal cellular walls.

## SC-CO$_2$: Reaction Media

Majority of chemical processes are carried out in organic solvents, which in most cases are toxic and flammable. Furthermore, these organic solvents need to be separated from the desired product and recycled back. To avoid these drawbacks, SCFs are suggested as an alternative.

As mentioned earlier, SCFs have gas-like diffusivities and low viscosities, which reduce mass resistance between reaction mixture and the catalyst and therefore result in an increase of reaction rate. Among possible solvents that can be used in supercritical conditions to conduct transesterification reactions, CO$_2$ was chosen due to its low critical temperature which make the process less energy intensive and more importantly below the denaturation temperature of the biocatalyst.

Kumar et al. [41] esterified palmitic acid with ethanol in temperature range of 35 to 70°C in the presence of three different lipases in SC-CO$_2$. Their results showed that Novozym 435 was the best catalyst. In SC-CO$_2$, Lipolase 100T and hog pancreas lipase showed similar results. Yields of 74, 44, and 40% were reached using Novozym 435, Lipolase 100T, and hog pancreas lipase, respectively, which were comparable to yield in solvent free system. Romero et al. [89] esterified isoamyl alcohol in SC-CO$_2$ and n-hexane. They noted that similar esterification degree was obtained in both SC-CO$_2$ and n-hexane systems; however, initial reaction rate was higher in SC-CO$_2$. Laudani et al. [218] compared FFA esterification with 1-octanol over immobilized lipase from R. miehei (Lipozyme RM IM) using three different reaction media: SC-CO$_2$, n-hexane, and solvent free systems. SC-CO$_2$ showed the highest conversion followed by n-hexane then solvent-free system.

Although SC-CO$_2$ has been used as a reaction media for enzyme esterification of FFA, limited work has been done on transesterification. D. Oliveira and J. V. Oliveira [219] compared enzymatic alcoholysis of palm kernel oil using n-hexane and SC-CO$_2$ systems. In SC-CO$_2$, highest conversion of 63% was obtained using Novozym 435 as catalyst whereas in n-hexane Lipozyme IM provided the highest conversion of

77%. Rathore and Madras [61] produced biodiesel from Jatropha oil with Novozym 435 in SC-CO$_2$. Optimum conditions were found to be 45°C, alcohol:oil molar ratio of 5:1, 30% enzyme loading, and 8 h with conversions of 60–70%. Varma and Madras [220] produced biodiesel from caster and linseed oils with Novozym 435 in SC-CO$_2$, and 45% yield in methanol and 35% in ethanol were obtained from linseed oil, whereas a very low yield of less than 10% was obtained from castor oil. Varma et al. [221] synthesized biodiesel from mustard and sesame oils using different acyl acceptors at 50°C for 24 h reaction. Their results showed that using mustard oil, conversion of roughly 70% and 65% can be obtained using methanol and ethanol, respectively. On the other hand, using sesame oil, a conversion of round 55% was obtained from ethanol, whereas only 45% conversion was obtained with methanol.

Despite the advantages of using SC-CO$_2$ as a reaction media for the enzymatic production of biodiesel, it has not been reported in any previous work on microalgae oil.

# MICROALGAE FOR BIODIESEL ENZYMATIC PRODUCTION USING SC-CO$_2$

## Microalgae as a Feedstock in Conventional Process

Recently, feasibility of using microalgae to produce biodiesel, as an alternative to fossil fuels, received significant attention since they are rich in lipids. Species like C. vulgaris, C. emersonii, Nannochloropsis sp.,P. tricornutum, and T. suecica have been reported in the literature for biodiesel production, where most of them were cultivated using glucose as a carbon source. However, glucose can be fermented directly to produce bioethanol.

Conventionally, microalgae have been used for biodiesel production using chemical catalytic reactions. Miao and Wu [120] studied biodiesel production from heterotrophic cultivated microalgae oil from C. protothecoides by 100% $H_2SO_4$ (based on oil weight) acidic transesterification. Biodiesel optimum conversion yield of 63% was obtained with 56:1 methanol:oil molar ratio at 30°C in 4 h reaction time. To overcome disadvantages of homogeneous catalysts, Carrero et al. [222] tested the ability of using hierarchical zeolites as heterogeneous catalyst.

With the target to reduce biodiesel production cost associated with oil extraction cost, in-situ transesterification, which is a direct conversion without solvent extraction, of the biomass oil to biodiesel has been performed [223, 224]. A conversion of 91% was achieved after 8 h of reaction at 60°C from Chlorella sp. [223] and 39, 40, 77, 78, and 82% were obtained from Synechocystis sp. PCC 6803Synechococcus elongates, Chlorella sorokiniana, T. suecica, and Chaetoceros gracilis at 80°C [224]. The high conversion obtained with Chlorella species could be due to the use of a stirring reactor that enhanced the mixing and reduced mass transfer resistances.

# Enzymatic Biodiesel Production from Microalgae

Similar to conventional feedstocks conversion, microalgae oil can also be converted to biodiesel using lipase. In this perspective, C. protothecoides is the only species that has been tested so far. Xiong et al. [225] produced biodiesel with 98% conversion from C. protothecoides with 30 wt% of Candida sp. lipase. Reaction conditions were 3:1 methanol:oil molar ratio, 10% water content, 38°C, pH of 7, and 12 h reaction. Similar conversion was obtained by Li et al. [154] at similar conditions but using 75 wt% of the immobilized lipase rather than 30 wt%. C. protothecoides was cultivated heterotrophically in both studies using glucose.

# Enzymatic Production with Sc-Co$_2$ Reaction Medium

To overcome the lipase inhibition limitations, mainly by methanol and glycerol, chemical solvents that can dissolve both methanol and glycerol have the advantage of increasing conversion yield. However, the use of organic solvents is not recommended due to its harmful environmental input and the solvent extraction unit.

Using $SC\text{-}CO_2$ as a reaction media adds to the advantages of organic solvents in saving downstream processing cost where product purification is not necessary. Since solubility is greatly influenced by fluid temperature and pressure adjustments, separation can be easily achieved by a pressure reduction where the product and enzyme do not dissolve at room temperature.

Due to its advantages over conventional organic solvents, the application of the high cost $SC\text{-}CO_2$ process may be justified in oil extraction from microalgae. However, its justification for biodiesel production may not be evident, despite its positive effect on reducing inhibition effects and easy product separation. Nevertheless, a combined continuous process of extracting oil from microalgae using $SC\text{-}CO_2$ and the use of the extracted oil for biodiesel production using immobilized lipase in $SC\text{-}CO_2$ in a one integrated system would economically be feasible. In this continuous process, the oil that is extracted from microalgae is already dissolved in $SC\text{-}CO_2$ and can be fed directly to the enzymatic bioreactor to produce biodiesel without the need for further expensive pumping. In this way, the attractive advantages of performing the reaction in $SC\text{-}CO_2$ media will be gained, avoiding at the same time the disadvantage of high pumping cost. Besides, using high pressure $CO_2$ might not have significant negative effect of lipase stability. Lanza et al. [226] investigated the influence of $SC\text{-}CO_2$ pressure on lipase activity and reported that the residual activity of Novozym-435 was approximately 90%. Previous study of D. Oliveira and J. V. Oliveira [219] on converting palm kernel oil to biodiesel using Novozym 435 showed that the rise in pressure in the range 60–90 bar actually results in an enhancement of initial reaction rate and conversion. However, at pressures beyond but 200 bar, a change in lipase structure may occur, which has a negative effect on the reaction. Therefore, the application of $SC\text{-}CO_2$ in the enzymatic reaction system should not exceed 200 bar.

# CONCLUSIONS

As verified in this paper, biodiesel produced from microalgae can realistically satisfy the global demand of diesel-fuel requirements. However, for cost effective production, this will not be applicable without microalgae biology and production processes enhancements. The potential of microalgal biodiesel production depends on selected microalgal strain and its ability to live in saline or use wastewaters and utilize $CO_2$ as a sole carbon source. In addition, biomass recovery that usually requires high energy and oil extraction has to be optimized for effective low cost overall process production. Another important point to be taken into consideration is the ability to use spent biomass, after oil extraction, for the production of other valuable coproducts such as animal feed or fertilizers.

The paper presents SC-$CO_2$ as a promising oil extraction technique from microalgae, and lipase as a biocatalyst for biodiesel production instead of the conventional chemical catalysts that require feed purification. The use of SC-$CO_2$ as a reaction media for the enzymatic production of biodiesel has also been discussed in the paper. Authors of this review, suggest future work to be done on designing an integrated SC-$CO_2$ extraction/reaction process, whereby a stream of extracted oil-rich SC-$CO_2$ from selected microalgae species is fed to a bioreactor containing lipase for enzymatic conversion of the oil into biodiesel.

# REFERENCES

1.  A. K. Agarwal, "Biofuels (alcohols and biodiesel) applications as fuels for internal combustion engines," Progress in Energy and Combustion Science, vol. 33, no. 3, pp. 233–271, 2007.

2.  E. G. Shay, "Diesel fuel from vegetable oils: status and opportunities," Biomass and Bioenergy, vol. 4, no. 4, pp. 227–242, 1993

3.  G. R. Peterson and W. P. Scarrah, "Rapeseed oil transesterification by heterogeneous catalysis," Journal of the American Oil Chemists Society, vol. 61, no. 10, pp. 1593–1597, 1984

4.  A. Demirbas, "Biodiesel from waste cooking oil via base-catalytic and supercritical methanol transesterification," Energy Conversion and Management, vol. 50, no. 4, pp. 923–927, 2009

5.  H. Fukuda, A. Kondo, and H. Noda, "Biodiesel fuel production by transesterification of oils," Journal of Bioscience and Bioengineering, vol. 92, no. 5, pp. 405–416, 2001.

6.  Y. Zhang, M. A. Dubé, D. D. McLean, and M. Kates, "Biodiesel production from waste cooking oil: 1. Process design and technological assessment," Bioresource Technology, vol. 89, no. 1, pp. 1–16, 2003.

7.  S. Al-Zuhair, "Production of biodiesel: possibilities and challenges," Biofuels, Bioproducts and Biorefining, vol. 1, no. 1, pp. 57–66, 2007

8.  F. Ma and M. A. Hanna, "Biodiesel production: a review," Bioresource Technology, vol. 70, no. 1, pp. 1–15, 1999.

9.  C. C. Akoh, S. W. Chang, G. C. Lee, and J. F. Shaw, "Enzymatic approach to biodiesel production,"Journal of Agricultural and Food Chemistry, vol. 55, no. 22, pp. 8995–9005, 2007

10. L. C. Meher, D. Vidya Sagar, and S. N. Naik, "Technical aspects of biodiesel production by transesterification—a review," Renewable and Sustainable Energy Reviews, vol. 10, no. 3, pp. 248–268, 2006.

11. Z. Helwani, M. R. Othman, N. Aziz, W. J. N. Fernando, and J. Kim, "Technologies for production of biodiesel focusing on green catalytic techniques: a review," Fuel Processing Technology, vol. 90, no. 12, pp. 1502–1514, 2009.

12. B. Freedman, E. H. Pryde, and T. L. Mounts, "Variables affecting the yields of fatty esters from transesterified vegetable oils," Journal of the American Oil Chemists Society, vol. 61, no. 10, pp. 1638–1643, 1984.

13. A. Demirbas, "Progress and recent trends in biofuels," Progress in Energy and Combustion Science, vol. 33, no. 1, pp. 1–18, 2007.

14. L. G. Schumacher, W. Marshall, J. Krahl, W. B. Wetherell, and M. S. Grabowski, "Biodiesel emissions data from series 60 DDC engines," Transactions of the American Society of Agricultural Engineers, vol. 44, no. 6, pp. 1465–1468, 2001. View at Scopus

15. D. Darnoko and M. Cheryan, "Kinetics of palm oil

transesterification in a batch reactor," Journal of the American Oil Chemists' Society, vol. 77, no. 12, pp. 1263–1267, 2000. View at Scopus

16. T. F. Dossin, M. F. Reyniers, R. J. Berger, and G. B. Marin, "Simulation of heterogeneously MgO-catalyzed transesterification for fine-chemical and biodiesel industrial production," Applied Catalysis B, vol. 67, no. 1-2, pp. 136–148, 2006.

17. B. Freedman, R. O. Butterfield, and E. H. Pryde, "Transesterification kinetics of soybean oil 1,"Journal of the American Oil Chemists' Society, vol. 63, no. 10, pp. 1375–1380, 1986.

18. W. Xie, H. Peng, and L. Chen, "Calcined Mg-Al hydrotalcites as solid base catalysts for methanolysis of soybean oil," Journal of Molecular Catalysis B, vol. 246, no. 1-2, pp. 24–32, 2006.

19. H. N. Bhatti, M. A. Hanif, M. Qasim, and Ata-ur-Rehman, "Biodiesel production from waste tallow,"Fuel, vol. 87, no. 13-14, pp. 2961–2966, 2008.

20. M. Canakci and J. Van Gerpen, "Biodiesel production from oils and fats with high free fatty acids,"Transactions of the American Society of Agricultural Engineers, vol. 44, no. 6, pp. 1429–1436, 2001.View at Scopus

21. C. M. Garcia, S. Teixeira, L. L. Marciniuk, and U. Schuchardt, "Transesterification of soybean oil catalyzed by sulfated zirconia," Bioresource Technology, vol. 99, no. 14, pp. 6608–6613, 2008.

22. M. K. Lam, K. T. Lee, and A. R. Mohamed, "Sulfated tin oxide as solid superacid catalyst for transesterification of waste cooking oil: an optimization study," Applied Catalysis B, vol. 93, no. 1-2, pp. 134–139, 2009.

23. S. Zheng, M. Kates, M. A. Dubé, and D. D. McLean, "Acid-catalyzed production of biodiesel from waste frying oil," Biomass and Bioenergy, vol. 30, no. 3, pp. 267–272, 2006.

24. W. Du, Y. Xu, D. Liu, and J. Zeng, "Comparative study on lipase-catalyzed transformation of soybean oil for biodiesel production with different acyl acceptors," Journal of Molecular Catalysis B, vol. 30, no. 3-4, pp. 125–129, 2004.

25. A.-F. Hsu, K. Jones, W. N. Marmer, and T. A. Foglia, "Production of alkyl esters from tallow and grease using lipase immobilized

in a phyllosilicate sol-gel," Journal of the American Oil Chemists' Society, vol. 78, no. 6, pp. 585–588, 2001. View at Scopus

26. M. K. Modi, J. R. C. Reddy, B. V. S. K. Rao, and R. B. N. Prasad, "Lipase-mediated conversion of vegetable oils into biodiesel using ethyl acetate as acyl acceptor," Bioresource Technology, vol. 98, no. 6, pp. 1260–1264, 2007.

27. H. Noureddini, X. Gao, and R. S. Philkana, "Immobilized Pseudomonas cepacia lipase for biodiesel fuel production from soybean oil," Bioresource Technology, vol. 96, no. 7, pp. 769–777, 2005.

28. O. Orçaire, P. Buisson, and A. C. Pierre, "Application of silica aerogel encapsulated lipases in the synthesis of biodiesel by transesterification reactions," Journal of Molecular Catalysis B, vol. 42, no. 3-4, pp. 106–113, 2006

29. S. Shah and M. N. Gupta, "Lipase catalyzed preparation of biodiesel from Jatropha oil in a solvent free system," Process Biochemistry, vol. 42, no. 3, pp. 409–414, 2007.

30. S. Saka and D. Kusdiana, "Biodiesel fuel from rapeseed oil as prepared in supercritical methanol,"Fuel, vol. 80, no. 2, pp. 225–231, 2001.

31. G. Madras, C. Kolluru, and R. Kumar, "Synthesis of biodiesel in supercritical fluids," Fuel, vol. 83, no. 14-15, pp. 2029–2033, 2004.

32. J. M. N. van Kasteren and A. P. Nisworo, "A process model to estimate the cost of industrial scale biodiesel production from waste cooking oil by supercritical transesterification," Resources, Conservation and Recycling, vol. 50, no. 4, pp. 442–458, 2007.

33. L. A. Nelson, T. A. Foglia, and W. N. Marmer, "Lipase-catalyzed production of biodiesel," Journal of the American Oil Chemists' Society, vol. 73, no. 9, pp. 1191–1195, 1996. View at Scopus

34. R. D. Abigor, P. O. Uadia, T. A. Foglia et al., "Lipase-catalysed production of biodiesel fuel from some Nigerian lauric oils," Biochemical Society Transactions, vol. 28, no. 6, pp. 979–981, 2000.

35. M. Kaieda, T. Samukawa, A. Kondo, and H. Fukuda, "Effect of methanol and water contents on production of biodiesel fuel from plant oil catalyzed by various lipases in a solvent-free

system,"Journal of Bioscience and Bioengineering, vol. 91, no. 1, pp. 12–15, 2001.

36.  C. J. Shieh, H. F. Liao, and C. C. Lee, "Optimization of lipase-catalyzed biodiesel by response surface methodology," Bioresource Technology, vol. 88, no. 2, pp. 103–106, 2003.

37.  M. Mittelbach, "Lipase catalyzed alcoholysis of sunflower oil," Journal of the American Oil Chemists' Society, vol. 67, no. 3, pp. 168–170, 1990.

38.  K. R. Jegannathan, S. Abang, D. Poncelet, E. S. Chan, and P. Ravindra, "Production of biodiesel using immobilized lipase—a critical review," Critical Reviews in Biotechnology, vol. 28, no. 4, pp. 253–264, 2008.

39.  L. Fjerbaek, K. V. Christensen, and B. Norddahl, "A review of the current state of biodiesel production using enzymatic transesterification," Biotechnology and Bioengineering, vol. 102, no. 5, pp. 1298–1315, 2009.

40.  M. Iso, B. Chen, M. Eguchi, T. Kudo, and S. Shrestha, "Production of biodiesel fuel from triglycerides and alcohol using immobilized lipase," Journal of Molecular Catalysis B, vol. 16, no. 1, pp. 53–58, 2001.

41.  R. Kumar, G. Madras, and J. Modak, "Enzymatic synthesis of ethyl palmitate in supercritical carbon dioxide," Industrial and Engineering Chemistry Research, vol. 43, no. 7, pp. 1568–1573, 2004. View at Scopus

42.  M. G. De Paola, E. Ricca, V. Calabrò, S. Curcio, and G. Iorio, "Factor analysis of transesterification reaction of waste oil for biodiesel production," Bioresource Technology, vol. 100, no. 21, pp. 5126–5131, 2009.

43.  M. Mukhopadhyay, Natural Extracts Using Supercritical Carbon Dioxide, CRC Press, New York, NY, USA, 2000.

44.  W. M. Antunes, C. D. O. Veloso, and C. A. Henriques, "Transesterification of soybean oil with methanol catalyzed by basic solids," Catalysis Today, vol. 133-135, no. 1-4, pp. 548–554, 2008.

45.  M. A. Kalam and H. H. Masjuki, "Biodiesel from palmoil—an analysis of its properties and potential,"Biomass and Bioenergy, vol. 23, no. 6, pp. 471–479, 2002.

46. A. N. Phan and T. M. Phan, "Biodiesel production from waste cooking oils," Fuel, vol. 87, no. 17-18, pp. 3490–3496, 2008.

47. A. Srivastava and R. Prasad, "Triglycerides-based diesel fuels," Renewable & Sustainable Energy Reviews, vol. 4, no. 2, pp. 111–133, 2000.

48. D. Rutz and R. Janssen, BioFuel Technology Handbook, WIP Renewable Energies, 2007.

49. Y. Chisti, "Biodiesel from microalgae," Biotechnology Advances, vol. 25, no. 3, pp. 294–306, 2007.

50. J. Sheehan, et al., A Look Back at the U.S. Department of Energy's Aquatic Species Program: Biodiesel from Algae, National Renewable Energy Laboratory, 1998.

51. A. Demirbas, "Progress and recent trends in biodiesel fuels," Energy Conversion and Management, vol. 50, no. 1, pp. 14–34, 2009.

52. D. Bajpai and V. K. Tyagi, "Biodiesel: source, production, composition, properties and its benifits,"Journal of Oleo Science, vol. 55, no. 10, pp. 487–502, 2006.

53. P. Fellows, Food Processing Technology: Principles and Practice, Woodhead Publishing Limited and CRC Press LLC, 2000.

54. S. Al-Zuhair, F. W. Ling, and L. S. Jun, "Proposed kinetic mechanism of the production of biodiesel from palm oil using lipase," Process Biochemistry, vol. 42, no. 6, pp. 951–960, 2007.

55. A. Demirbas, "Importance of biodiesel as transportation fuel," Energy Policy, vol. 35, no. 9, pp. 4661–4670, 2007.

56. S. V. Ranganathan, S. L. Narasimhan, and K. Muthukumar, "An overview of enzymatic production of biodiesel," Bioresource Technology, vol. 99, no. 10, pp. 3975–3981, 2008

57. M. A. Hess, M. J. Haas, T. A. Foglia, and W. N. Marmer, "Effect of antioxidant addition on NOx emissions from biodiesel," Energy & Fuels, vol. 19, no. 4, pp. 1749–1754, 2005.

58. W. G. Wang, D. W. Lyons, N. N. Clark, M. Gautam, and P. M. Norton, "Emissions from nine heavy trucks fueled by diesel and biodiesel blend without engine modification," Environmental Science & Technology, vol. 34, no. 6, pp. 933–939, 2000.

59. S. A. Basha, K. R. Gopal, and S. Jebaraj, "A review on biodiesel production, combustion, emissions and performance,"

Renewable and Sustainable Energy Reviews, vol. 13, no. 6-7, pp. 1628–1634, 2009.

60. Y. C. Sharma, B. Singh, and S. N. Upadhyay, "Advancements in development and characterization of biodiesel: a review," Fuel, vol. 87, no. 12, pp. 2355–2377, 2008.

61. V. Rathore and G. Madras, "Synthesis of biodiesel from edible and non-edible oils in supercritical alcohols and enzymatic synthesis in supercritical carbon dioxide," Fuel, vol. 86, no. 17-18, pp. 2650–2659, 2007.

62. A. Robles-Medina, P. A. González-Moreno, L. Esteban-Cerdán, and E. Molina-Grima, "Biocatalysis: towards ever greener biodiesel production," Biotechnology Advances, vol. 27, no. 4, pp. 398–408, 2009.

63. X. Fan and R. Burton, "Recent development of biodiesel feedstocks and the applications of glycerol: a review," The Open Fuels & Energy Science Journal, vol. 1, pp. 100–109, 2009.

64. M. A. Dubé, A. Y. Tremblay, and J. Liu, "Biodiesel production using a membrane reactor,"Bioresource Technology, vol. 98, no. 3, pp. 639–647, 2007.

65. J. Van Gerpen, "Biodiesel processing and production," Fuel Processing Technology, vol. 86, no. 10, pp. 1097–1107, 2005

66. J. M. Marchetti, V. U. Miguel, and A. F. Errazu, "Possible methods for biodiesel production,"Renewable and Sustainable Energy Reviews, vol. 11, no. 6, pp. 1300–1311, 2007.

67. S. Sinha, A. K. Agarwal, and S. Garg, "Biodiesel development from rice bran oil: transesterification process optimization and fuel characterization," Energy Conversion and Management, vol. 49, no. 5, pp. 1248–1257, 2008.

68. K. Rajan and K. R. Senthilkumar, "Effect of exhaust gas recirculation (EGR) on the performance and emission characteristics of diesel engine with sunflower oil methyl ester," Jordan Journal of Mechanical and Industrial Engineering, vol. 3, no. 4, pp. 306–311, 2009.

69. L. Lin, D. Ying, S. Chaitep, and S. Vittayapadung, "Biodiesel production from crude rice bran oil and properties as fuel," Applied Energy, vol. 86, no. 5, pp. 681–688, 2009.

70. F. Ma, L. D. Clements, and M. A. Hanna, "Biodiesel fuel from animal fat. Ancillary studies on transesterification of beef tallow," Industrial and Engineering Chemistry Research, vol. 37, no. 9, pp. 3768–3771, 1998. View at Scopus

71. A. Sivasamy, K. Y. Cheah, P. Fornasiero, F. Kemausuor, S. Zinoviev, and S. Miertus, "Catalytic applications in the production of biodiesel from vegetable oils," ChemSusChem, vol. 2, no. 4, pp. 278–300, 2009.

72. M. L. Granados, M. D. Z. Poves, D. M. Alonso et al., "Biodiesel from sunflower oil by using activated calcium oxide," Applied Catalysis B, vol. 73, no. 3, pp. 317–326, 2007.

73. T. Ebiura, T. Echizen, A. Ishikawa, K. Murai, and T. Baba, "Selective transesterification of triolein with methanol to methyl oleate and glycerol using alumina loaded with alkali metal salt as a solid-base catalyst," Applied Catalysis A, vol. 283, no. 1-2, pp. 111–116, 2005

74. C. C. C. M. Silva, N. F. P. Ribeiro, M. M. V. M. Souza, and D. A. G. Aranda, "Biodiesel production from soybean oil and methanol using hydrotalcites as catalyst," Fuel Processing Technology, vol. 91, no. 2, pp. 205–210, 2010

75. S. Zhang, Y. G. Zu, Y. J. Fu, M. Luo, D. Y. Zhang, and T. Efferth, "Rapid microwave-assisted transesterification of yellow horn oil to biodiesel using a heteropolyacid solid catalyst," Bioresource Technology, vol. 101, no. 3, pp. 931–936, 2010.

76. N. Shibasaki-Kitakawa, H. Honda, H. Kuribayashi, T. Toda, T. Fukumura, and T. Yonemoto, "Biodiesel production using anionic ion-exchange resin as heterogeneous catalyst," Bioresource Technology, vol. 98, no. 2, pp. 416–421, 2007.

77. A. Demirba , "Biodiesel from vegetable oils via transesterification in supercritical methanol," Energy Conversion and Management, vol. 43, no. 17, pp. 2349–2356, 2002

78. K. T. Tan, K. T. Lee, and A. R. Mohamed, "Production of FAME by palm oil transesterification via supercritical methanol technology," Biomass and Bioenergy, vol. 33, no. 8, pp. 1096–1099, 2009.

79. M. J. Haas and T. A. Foglia, "Alternate feedstocks and technologies for biodiesel production," in The Biodiesel Handbook, G. Knothe, J. V. Gerpen, and J. Krahl, Eds., pp. 42–61, AOCS Press, Champaign, Ill, USA, 2005.

80. L. Li, W. Du, D. Liu, L. Wang, and Z. Li, "Lipase-catalyzed transesterification of rapeseed oils for biodiesel production with a novel organic solvent as the reaction medium," Journal of Molecular Catalysis B, vol. 43, no. 1-4, pp. 58–62, 2006.

81. M. Szczesna Antczak, A. Kubiak, T. Antczak, and S. Bielecki, "Enzymatic biodiesel synthesis—key factors affecting efficiency of the process," Renewable Energy, vol. 34, no. 5, pp. 1185–1194, 2009.

82. K. R. Jegannathan, et al., "Design an immobilized lipase enzyme for biodiesel production," Journal of Renewable Sustainable Energy, vol. 1, no. 6, p. 063101, 2009.

83. V. Caballero, F. M. Bautista, J. M. Campelo et al., "Sustainable preparation of a novel glycerol-free biofuel by using pig pancreatic lipase: partial 1,3-regiospecific alcoholysis of sunflower oil," Process Biochemistry, vol. 44, no. 3, pp. 334–342, 2009.

84. L. Cao, "Immobilised enzymes: science or art?" Current Opinion in Chemical Biology, vol. 9, no. 2, pp. 217–226, 2005.

85. C. M. Drapcho, N. P. Nhuan, and T. H. Walker, Biofuels Engineering Process Technology, McGraw-Hill Companies, 2008.

86. Ö. Köse, M. Tüter, and H. A. Aksoy, "Immobilized Candida antarctica lipase-catalyzed alcoholysis of cotton seed oil in a solvent-free medium," Bioresource Technology, vol. 83, no. 2, pp. 125–129, 2002.

87. D. Royon, M. Daz, G. Ellenrieder, and S. Locatelli, "Enzymatic production of biodiesel from cotton seed oil using t-butanol as a solvent," Bioresource Technology, vol. 98, no. 3, pp. 648–653, 2007

88. L. Wang, W. Du, D. Liu, L. Li, and N. Dai, "Lipase-catalyzed biodiesel production from soybean oil deodorizer distillate with absorbent present in tert-butanol system," Journal of Molecular Catalysis B, vol. 43, no. 1-4, pp. 29–32, 2006. View at Publisher · View at Google Scholar · View at Scopus

89. M. D. Romero, et al., "Enzymatic synthesis of isoamyl acetate with immobilized Candida antarctica lipase in supercritical carbon dioxide," Journal of Supercritical Fluids, vol. 33, no. 1, pp. 77–84, 2005.

90.  X. Liu, H. He, Y. Wang, S. Zhu, and X. Piao, "Transesterification of soybean oil to biodiesel using CaO as a solid base catalyst," Fuel, vol. 87, no. 2, pp. 216–221, 2008.

91.  A. Z. Abdullah, B. Salamatinia, H. Mootabadi, and S. Bhatia, "Current status and policies on biodiesel industry in Malaysia as the world›s leading producer of palm oil," Energy Policy, vol. 37, no. 12, pp. 5440–5448, 2009.

92.  G. A. Pereyra-Irujo, N. G. Izquierdo, M. Covi, S. M. Nolasco, F. Quiroz, and L. A. N. Aguirrezábal, "Variability in sunflower oil quality for biodiesel production: a simulation study," Biomass and Bioenergy, vol. 33, no. 3, pp. 459–468, 2009.

93.  K. G. Georgogianni, M. G. Kontominas, P. J. Pomonis, D. Avlonitis, and V. Gergis, "Conventional and in situ transesterification of sunflower seed oil for the production of biodiesel," Fuel Processing Technology, vol. 89, no. 5, pp. 503–509, 2008.

94.  Z. Wen, X. Yu, S. T. Tu, J. Yan, and E. Dahlquist, "Biodiesel production from waste cooking oil catalyzed by $TiO_2$-MgO mixed oxides," Bioresource Technology, vol. 101, no. 24, pp. 9570–9576, 2010.

95.  P. Patil, S. Deng, J. Isaac Rhodes, and P. J. Lammers, "Conversion of waste cooking oil to biodiesel using ferric sulfate and supercritical methanol processes," Fuel, vol. 89, no. 2, pp. 360–364, 2010

96.  Y. Wang, S. O. Pengzhan Liu, and Z. Zhang, "Preparation of biodiesel from waste cooking oil via two-step catalyzed process," Energy Conversion and Management, vol. 48, no. 1, pp. 184–188, 2007.

97.  D. Y. C. Leung, X. Wu, and M. K. H. Leung, "A review on biodiesel production using catalyzed transesterification," Applied Energy, vol. 87, no. 4, pp. 1083–1095, 2010.

98.  H. Taher, et al., "Extracted fat from lamb meat by supercritical $CO_2$ as feedstock for biodiesel production," Biochemical Engineering Journal, vol. 55, no. 1, pp. 23–31, 2011.

99.  S. Behzadi and M. M. Farid, "Production of biodiesel using a continuous gas-liquid reactor,"Bioresource Technology, vol. 100, no. 2, pp. 683–689, 2009.

100. S. T. Hoh and M. Farid, "Examining glycerolysis as a pre-treatment method for low quality biodiesel feedstock," in Proceedings of the Chemeca, Adelaide, Australia, September 2010.

101. M. Adamczak, U. T. Bornscheuer, and W. Bednarski, "The application of biotechnological methods for the synthesis of biodiesel," European Journal of Lipid Science and Technology, vol. 111, no. 8, pp. 800–813, 2009.

102. A. Demirbas, "Use of algae as biofuel sources," Energy Conversion and Management, vol. 51, no. 12, pp. 2738–2749, 2010

103. A. Demirbas and M. Fatih Demirbas, "Importance of algae oil as a source of biodiesel," Energy Conversion and Management, vol. 52, no. 1, pp. 163–170, 2011.

104. L. E. Graham and L. W. Wilcox, Algae, Prentice Hall, Upper Saddle River, NJ, USA, 2000.

105. G. Pokoo-Aikins, A. Nadim, M. M. El-Halwagi, and V. Mahalec, "Design and analysis of biodiesel production from algae grown through carbon sequestration," Clean Technologies and Environmental Policy, vol. 12, no. 3, pp. 239–254, 2010.

106. A. P. Vyas, J. L. Verma, and N. Subrahmanyam, "A review on FAME production processes," Fuel, vol. 89, no. 1, pp. 1–9, 2010.

107. R. E. Lee, Phycology, Cambridge University Press, New York, NY, USA, 4th edition, 2008.

108. L. Brennan and P. Owende, "Biofuels from microalgae—a review of technologies for production, processing, and extractions of biofuels and co-products," Renewable and Sustainable Energy Reviews, vol. 14, no. 2, pp. 557–577, 2010.

109. Y.-K. Lee, "Microalgal mass culture systems and methods: their limitation and potential," Journal of Applied Phycology, vol. 13, no. 4, pp. 307–315, 2001.

110. J. N. Rosenberg, G. A. Oyler, L. Wilkinson, and M. J. Betenbaugh, "A green light for engineered algae: redirecting metabolism to fuel a biotechnology revolution," Current Opinion in Biotechnology, vol. 19, no. 5, pp. 430–436, 2008.

111. V. Patil, K.-Q. Tran, and H. R. Giselrød, "Towards sustainable production of biofuels from microalgae," International Journal of Molecular Sciences, vol. 9, no. 7, pp. 1188–1195, 2008

112. K. W. Crane and J. P. Grover, "Coexistence of mixotrophs, autotrophs, and heterotrophs in planktonic microbial communities," Journal of Theoretical Biology, vol. 262, no. 3, pp. 517–527, 2010.

113. L. Barsanti and P. Gualtieri, Algae: Anatomy, Biochemistry, and Biotechnology, Taylor & Francis Group, CRC Press, New York, NY, USA, 2006.

114. L. Gouveia and A. C. Oliveira, "Microalgae as a raw material for biofuels production," Journal of Industrial Microbiology & Biotechnology, vol. 36, no. 2, pp. 269–274, 2009.

115. W. Becker, "Microalgae in human and animal nutrition," in Handbook of Microalgal Culture: Biotechnology and Applied Phycology, A. Richmond, Ed., pp. 312–351, Blackwell Science, 2004.

116. P. Sze, A Biology of the Algae, vol. 3rd, McGraw-Hill Science, 1998.

117. D. Voltolina, M. D. P. Sánchez-Saavedra, and L. M. Torres-Rodríguez, "Outdoor mass microalgae production in Bahia Kino, Sonora, NW Mexico," Aquacultural Engineering, vol. 38, no. 2, pp. 93–96, 2008.

118. F. M. I. Natrah, F. M. Yusoff, M. Shariff, F. Abas, and N. S. Mariana, "Screening of Malaysian indigenous microalgae for antioxidant properties and nutritional value," Journal of Applied Phycology, vol. 19, no. 6, pp. 711–718, 2007.

119. S. M. Renaud, L. V. Thinh, G. Lambrinidis, and D. L. Parry, "Effect of temperature on growth, chemical composition and fatty acid composition of tropical Australian microalgae grown in batch cultures," Aquaculture, vol. 211, no. 1-4, pp. 195–214, 2002

120. X. Miao and Q. Wu, "Biodiesel production from heterotrophic microalgal oil," Bioresource Technology, vol. 97, no. 6, pp. 841–846, 2006.

121. X. Miao and Q. Wu, "High yield bio-oil production from fast pyrolysis by metabolic controlling of Chlorella protothecoides," Journal of Biotechnology, vol. 110, no. 1, pp. 85–93, 2004.

122. X. Miao, Q. Wu, and C. Yang, "Fast pyrolysis of microalgae to produce renewable fuels," Journal of Analytical and Applied Pyrolysis, vol. 71, no. 2, pp. 855–863, 2004.

123. J. Fábregas, A. Maseda, A. Domínguez, and A. Otero, "The cell composition of Nannochloropsis sp. changes under different irradiances in semicontinuous culture," World Journal of

Microbiology and Biotechnology, vol. 20, no. 1, pp. 31–35, 2004.

124. S. Repka, M. Van Der Vlies, and J. Vijverberg, "Food quality of detritus derived from the filamentous cyanobacterium Oscillatoria limnetica for Daphnia galeata," Journal of Plankton Research, vol. 20, no. 11, pp. 2199–2205, 1998.

125. Y. Singh and H. D. Kumar, "Lipid and hydrocarbon production by Botryococcus spp. under nitrogen limitation and anaerobiosis," World Journal of Microbiology & Biotechnology, vol. 8, no. 2, pp. 121–124, 1992.

126. L. Rodolfi, G. C. Zittelli, N. Bassi et al., "Microalgae for oil: strain selection, induction of lipid synthesis and outdoor mass cultivation in a low-cost photobioreactor," Biotechnology and Bioengineering, vol. 102, no. 1, pp. 100–112, 2009.

127. J. R. Dean, Extraction Methods for Environmental Analysis, John Wiley & Sons, Chichester, UK, 1998.

128. R. L. Mendes, H. L. Fernandes, J. P. Coelho et al., "Supercritical $CO_2$ extraction of carotenoids and other lipids from Chlorella vulgaris," Food Chemistry, vol. 53, no. 1, pp. 99–103, 1995.

129. G. Andrich, U. Nesti, F. Venturi, A. Zinnai, and R. Fiorentini, "Supercritical fluid extraction of bioactive lipids from the microalga Nannochloropsis sp," European Journal of Lipid Science and Technology, vol. 107, no. 6, pp. 381–386, 2005

130. G. Andrich, A. Zinnai, U. Nesti, F. Venturi, and R. Fiorentini, "Supercritical fluid extraction of oil from microalga Spirulina (Arthrospira) platensis," Acta Alimentaria, vol. 35, no. 2, pp. 195–203, 2006.

131. R. Halim, B. Gladman, M. K. Danquah, and P. A. Webley, "Oil extraction from microalgae for biodiesel production," Bioresource Technology, vol. 102, no. 1, pp. 178–185, 2011

132. R. L. Mendes, B. P. Nobre, M. T. Cardoso, A. P. Pereira, and A. F. Palavra, "Supercritical carbon dioxide extraction of compounds with pharmaceutical importance from microalgae," Inorganica Chimica Acta, vol. 356, pp. 328–334, 2003

133. P. Spolaore, C. Joannis-Cassan, E. Duran, and A. Isambert, "Commercial applications of microalgae," Journal of Bioscience and Bioengineering, vol. 101, no. 2, pp. 87–96, 2006.

134. K. Vijayaraghavan and K. Hemanathan, "Biodiesel production from freshwater algae," Energy & Fuels, vol. 23, no. 11, pp. 5448–5453, 2009

135. X. Meng, J. Yang, X. Xu, L. Zhang, Q. Nie, and M. Xian, "Biodiesel production from oleaginous microorganisms," Renewable Energy, vol. 34, no. 1, pp. 1–5, 2009.

136. Q. Hu, M. Sommerfeld, E. Jarvis et al., "Microalgal triacylglycerols as feedstocks for biofuel production: perspectives and advances," Plant Journal, vol. 54, no. 4, pp. 621–639, 2008.

137. G. Huang, F. Chen, D. Wei, X. Zhang, and G. Chen, "Biodiesel production by microalgal biotechnology," Applied Energy, vol. 87, no. 1, pp. 38–46, 2010.

138. S. N. Naik, V. V. Goud, P. K. Rout, and A. K. Dalai, "Production of first and second generation biofuels: a comprehensive review," Renewable and Sustainable Energy Reviews, vol. 14, no. 2, pp. 578–597, 2010

139. S. A. Khan, Rashmi, M. Z. Hussain, S. Prasad, and U. C. Banerjee, "Prospects of biodiesel production from microalgae in India," Renewable and Sustainable Energy Reviews, vol. 13, no. 9, pp. 2361–2372, 2009

140. J. Knauer and P. C. Southgate, "A review of the nutritional requirements of bivalves and the development of alternative and artificial diets for bivalve aquaculture," Reviews in Fisheries Science, vol. 7, no. 3-4, pp. 241–280, 1999.

141. E. W. Becker, "Micro-algae as a source of protein," Biotechnology Advances, vol. 25, no. 2, pp. 207–210, 2007

142. J. L. Harwood and I. A. Guschina, "The versatility of algae and their lipid metabolism," Biochimie, vol. 91, no. 6, pp. 679–684, 2009

143. C. M. Beal, C. H. Smith, M. E. Webber, R. S. Ruoff, and R. E. Hebner, "A framework to report the production of renewable diesel from algae," Bioenergy Research, vol. 4, no. 1, pp. 36–60, 2011.

144. M. Gavrilescu and Y. Chisti, "Biotechnology—a sustainable alternative for chemical industry,"Biotechnology Advances, vol. 23, no. 7-8, pp. 471–499, 2005.

145. P. G. Roessler, et al., "Genetic-engineering approaches for enhanced production of biodiesel fuel from microalgae," in ACS Symposium Series, pp. 255–270, 1994.

146. J. Singh and S. Gu, "Commercialization potential of microalgae for biofuels production," Renewable and Sustainable Energy Reviews, vol. 14, pp. 2596–2610, 2010.

147. M. Aresta, A. Dibenedetto, M. Carone, T. Colonna, and C. Fragale, "Production of biodiesel from macroalgae by supercritical $CO_2$ extraction and thermochemical liquefaction," Environmental Chemistry Letters, vol. 3, no. 3, pp. 136–139, 2005.

148. K. G. Satyanarayana, A. B. Mariano, and J. V. C. Vargas, "A review on microalgae, a versatile source for sustainable energy and materials," International Journal of Energy Research, vol. 35, no. 4, pp. 291–311, 2011.

149. J. Singh and S. Gu, "Commercialization potential of microalgae for biofuels production," Renewable and Sustainable Energy Reviews, vol. 14, no. 9, pp. X2596–2610, 2010.

150. H. C. Greenwell, L. M. L. Laurens, R. J. Shields, R. W. Lovitt, and K. J. Flynn, "Placing microalgae on the biofuels priority list: a review of the technological challenges," Journal of the Royal Society Interface, vol. 7, no. 46, pp. 703–726, 2010.

151. T. M. Mata, A. A. Martins, and N. S. Caetano, "Microalgae for biodiesel production and other applications: a review," Renewable and Sustainable Energy Reviews, vol. 14, no. 1, pp. 217–232, 2010.

152. L. M. Brown and K. G. Zeiler, "Aquatic biomass and carbon dioxide trapping," Energy Conversion and Management, vol. 34, no. 9-11, pp. 1005–1013, 1993.

153. P. Schenk, et al., "Second generation biofuels: high-efficiency microalgae for biodiesel production," BioEnergy Research, vol. 1, no. 1, pp. 20–43, 2008.

154. X. Li, H. Xu, and Q. Wu, "Large-scale biodiesel production from microalga Chlorella protothecoides through heterotrophic cultivation in bioreactors," Biotechnology and Bioengineering, vol. 98, no. 4, pp. 764–771, 2007.

155. Y. Cheng, W. Zhou, C. Gao, K. Lan, Y. Gao, and Q. Wu, "Biodiesel production from Jerusalem artichoke (Helianthus

Tuberosus L.) tuber by heterotrophic microalgae Chlorella protothecoides,"Journal of Chemical Technology & Biotechnology, vol. 84, no. 5, pp. 777–781, 2009

156. Y. Lu, Y. Zhai, M. Liu, and Q. Wu, "Biodiesel production from algal oil using cassava (Manihot esculenta Crantz) as feedstock," Journal of Applied Phycology, vol. 22, no. 5, pp. 573–578, 2010

157. P. T. Vasudevan and M. Briggs, "Biodiesel production—current state of the art and challenges,"Journal of Industrial Microbiology & Biotechnology, vol. 35, no. 5, pp. 421–430, 2008

158. A. H. Scragg, Biofuels, Production, Application and Development, Cambridge University Press, Cambridge, UK, 2009.

159. R. A. Holser and R. Harry-O'Kuru, "Transesterified milkweed (Asclepias) seed oil as a biodiesel fuel,"Fuel, vol. 85, no. 14-15, pp. 2106–2110, 2006.

160. N. A. Santos, J. R. J. Santos, F. S. M. Sinfrônio et al., "Thermo-oxidative stability and cold flow properties of babassu biodiesel by PDSC and TMDSC techniques," Journal of Thermal Analysis and Calorimetry, vol. 97, no. 2, pp. 611–614, 2009.

161. A. Richmond, "Open systems for the mass production of photoautotrophic microalgae outdoors: physiological principles," Journal of Applied Phycology, vol. 4, no. 3, pp. 281–286, 1992

162. R. Harun, M. Singh, G. M. Forde, and M. K. Danquah, "Bioprocess engineering of microalgae to produce a variety of consumer products," Renewable and Sustainable Energy Reviews, vol. 14, no. 3, pp. 1037–1047, 2010.

163. M. R. Tredici and R. Materassi, "From open ponds to vertical alveolar panels: the Italian experience in the development of reactors for the mass cultivation of phototrophic microorganisms," Journal of Applied Phycology, vol. 4, no. 3, pp. 221–231, 1992

164. D. Chaumont, "Biotechnology of algal biomass production: a review of systems for outdoor mass culture," Journal of Applied Phycology, vol. 5, no. 6, pp. 593–604, 1993.

165. O. R. Zaborsky, et al., "An automated helical photobioreactor incorporating cyanobacteria for continuous hydrogen production," in BioHydrogen, pp. 431–440, Springer, New York, NY, USA, 1999.

166. P. T. Vasudevan and B. Fu, "Environmentally sustainable biofuels:

advances in biodiesel research,"Waste and Biomass Valorization, pp. 1–17, 2010.

167. M. E. Huntley and D. G. Redalje, "$CO_2$ mitigation and renewable oil from photosynthetic microbes: a new appraisal," Mitigation and Adaptation Strategies for Global Change, vol. 12, no. 4, pp. 573–608, 2007.

168. A. M. Illman, A. H. Scragg, and S. W. Shales, "Increase in Chlorella strains calorific values when grown in low nitrogen medium," Enzyme and Microbial Technology, vol. 27, no. 8, pp. 631–635, 2000.

169. A. H. Scragg, A. M. Illman, A. Carden, and S. W. Shales, "Growth of microalgae with increased calorific values in a tubular bioreactor," Biomass and Bioenergy, vol. 23, no. 1, pp. 67–73, 2002.

170. A. Widjaja, C. C. Chien, and Y. H. Ju, "Study of increasing lipid production from fresh water microalgae Chlorella vulgaris," Journal of the Taiwan Institute of Chemical Engineers, vol. 40, no. 1, pp. 13–20, 2009.

171. S.-Y. Chiu, C. Y. Kao, M. T. Tsai, S. C. Ong, C. H. Chen, and C. S. Lin, "Lipid accumulation and $CO_2$ utilization of Nannochloropsis oculata in response to $CO_2$ aeration," Bioresource Technology, vol. 100, no. 2, pp. 833–838, 2009

172. H. Tang, et al., "Potential of microalgae oil from Dunaliella tertiolecta as a feedstock for biodiesel,"Applied Energy, vol. 88, no. 10, pp. 3324–3330, 2011.

173. W. R. Barclay, K. M. Meager, and J. R. Abril, "Heterotrophic production of long chain omega-3 fatty acids utilizing algae and algae-like microorganisms," Journal of Applied Phycology, vol. 6, no. 2, pp. 123–129, 1994.

174. Y. Liang, N. Sarkany, and Y. Cui, "Biomass and lipid productivities of Chlorella vulgaris under autotrophic, heterotrophic and mixotrophic growth conditions," Biotechnology Letters, vol. 31, no. 7, pp. 1043–1049, 2009.

175. K. Muffler and R. Ulber, "Downstream processing in marine biotechnology," in Marine Biotechnology II, Y. L. Gal and R. Ulber, Eds., p. 261, Springer, New York, NY, USA, 1st edition, 2005.

176. J. O'Grady and J. A. Morgan, "Heterotrophic growth and lipid production of Chlorella protothecoides on glycerol," Bioprocess and Biosystems Engineering, vol. 34, no. 1, pp. 121–125, 2011.

177. H. Xu, X. Miao, and Q. Wu, "High quality biodiesel production from a microalga Chlorella protothecoides by heterotrophic growth in fermenters," Journal of Biotechnology, vol. 126, no. 4, pp. 499–507, 2006

178. J. Liu, J. Huang, Z. Sun, Y. Zhong, Y. Jiang, and F. Chen, "Differential lipid and fatty acid profiles of photoautotrophic and heterotrophic Chlorella zofingiensis: assessment of algal oils for biodiesel production," Bioresource Technology, vol. 102, no. 1, pp. 106–110, 2011.

179. L. E. Dong, K. J. Drury, and A. G. Fadeev, "Methods for Harvesting Biological Materials Using Membrane Filters," US Patent 20100184197, 2010.

180. A. Lamers, "Algae oils from small scale low input water remediation site as feedstock for biodiesel conversion," Guelph Engineering Journal, vol. 2, pp. 24–38, 2009.

181. Y. Li, M. Horsman, N. Wu, C. Q. Lan, and N. Dubois-Calero, "Biofuels from microalgae,"Biotechnology Progress, vol. 24, no. 4, pp. 815–820, 2008.

182. M. K. Danquah, L. Ang, N. Uduman, N. Moheimani, and G. M. Forde, "Dewatering of microalgal culture for biodiesel production: exploring polymer flocculation and tangential flow filtration,"Journal of Chemical Technology & Biotechnology, vol. 84, no. 7, pp. 1078–1083, 2009.

183. N. Uduman, et al., "Dewatering of microalgal cultures: a major bottleneck to algae-based fuels,"Journal of Renewable Sustainable Energy, vol. 2, p. 012701, 2010.

184. E. Molina Grima, E. H. Belarbi, F. G. Acién Fernández, A. Robles Medina, and Y. Chisti, "Recovery of microalgal biomass and metabolites: process options and economics," Biotechnology Advances, vol. 20, no. 7-8, pp. 491–515, 2003.

185. C. Gudin and C. Therpenier, "Bioconversion of solar energy into organic chemicals by microalgae,"Advanced Biotechnology Processes, vol. 6, pp. 73–110, 1986.

186. M. Olaizola, "Commercial development of microalgal

biotechnology: from the test tube to the marketplace," Biomolecular Engineering, vol. 20, no. 4-6, pp. 459–466, 2003

187. H. M. Oh, S. J. Lee, M. H. Park et al., "Harvesting of Chlorella vulgaris using a bioflocculant from Paenibacillus sp. AM49," Biotechnology Letters, vol. 23, no. 15, pp. 1229–1234, 2001.

188. M. W. Tenney, et al., "Algal flocculation with synthetic organic polyelectrolytes," Applied and Environmental Microbiology, vol. 18, no. 6, pp. 965–971, 1969.

189. G. Shelef, A. Sukenik, and M. Green, "Microalgae harvesting and processing: a literature review," Tech. Rep., Technion Research and Development, 1984.

190. G. Ahlgren and L. Merino, "Lipid analysis of freshwater microalgae : a method study," Archiv für Hydrobiologie, vol. 121, no. 3, pp. 295–306, 1991.

191. E. H. Belarbi, E. Molina, and Y. Chisti, "A process for high yield and scaleable recovery of high purity eicosapentaenoic acid esters from microalgae and fish oil," Process Biochemistry, vol. 35, no. 9, pp. 951–969, 2000.

192. E. M. Grima, A. R. Medina, A. G. Giménez, J. A. Sánchez Pérez, F. G. Camacho, and J. L. García Sánchez, "Comparison between extraction of lipids and fatty acids from microalgal biomass," Journal of the American Oil Chemists' Society, vol. 71, no. 9, pp. 955–959, 1994.

193. A. R. Fajardo, L. E. Cerdán, A. R. Medina, F. G. A. Fernández, P. A. G. Moreno, and E. M. Grima, "Lipid extraction from the microalga Phaeodactylum tricornutum," European Journal of Lipid Science and Technology, vol. 109, no. 2, pp. 120–126, 2007

194. C. Y. Chen, K. L. Yeh, H. M. Su, Y. C. Lo, W. M. Chen, and J. S. Chang, "Strategies to enhance cell growth and achieve high-level oil production of a Chlorella vulgaris isolate," Biotechnology Progress, vol. 26, no. 3, pp. 679–686, 2010.

195. A. Morist, J. L. Montesinos, J. A. Cusidó, and F. Gòdia, "Recovery and treatment of Spirulina platensis cells cultured in a continuous photobioreactor to be used as food," Process Biochemistry, vol. 37, no. 5, pp. 535–547, 2001.

196. E. A. Ehimen, et al., "Energy recovery from lipid extracted,

transesterified and glycerol codigested microalgae biomass," GCB Bioenergy, vol. 1, no. 6, pp. 371–381, 2009.

197. Y. Chisti and M. Moo-Young, "Disruption of microbial cells for intracellular products," Enzyme and Microbial Technology, vol. 8, no. 4, pp. 194–204, 1986.

198. M. Ottens, J. A. Wesselingh, and L. A. M. V. D. Wielen, "Downstream processing," in Basic Biotechnology, C. Ratledge and B. Kristiansen, Eds., p. 666, Cambridge University Press, New York, NY, USA, 3rd edition, 2006.

199. C. L. Peterson, D. L. Auld, and R. A. Korus, "Winter rape oil fuel for diesel engines: recovery and utilization," Journal of the American Oil Chemists' Society, vol. 60, no. 8, pp. 1579–1587, 1983.

200. Z. Khan, J. Troquet, and C. Vachelard, "Sample preparation and analytical techniques for determination of polyaromatic hydrocarbons in soils," International Journal of Environmental Science and Technology, vol. 2, no. 3, pp. 275–286, 2005.

201. E. G. Bligh and W. J. Dyer, "A rapid method of total lipid extraction and purification," Canadian Journal of Biochemistry and Physiology, vol. 37, no. 8, pp. 911–917, 1959.

202. S. Lee, B.-D. Yoon, and H.-. Oh, "Rapid method for the determination of lipid from the green alga Botryococcus braunii," Biotechnology Techniques, vol. 12, no. 7, pp. 553–556, 1998.

203. N. Nagle and P. Lemke, "Production of methyl ester fuel from microalgae," Applied Biochemistry and Biotechnology, vol. 24-25, pp. 355–361, 1990.

204. L. Govindarajan, N. Raut, and A. Alsaeed, "Novel solvent extraction for extraction of oil from algae biomass growth in desalination reject stream," Journal of Algal Biomass Utilization, vol. 1, no. 1, pp. 18–28, 2009.

205. M. Herrero, A. Cifuentes, and E. Ibañez, "Sub- and supercritical fluid extraction of functional ingredients from different natural sources: plants, food-by-products, algae and microalgae—a review," Food Chemistry, vol. 98, no. 1, pp. 136–148, 2006.

206. M. Bender, "Economic feasibility review for community-scale farmer cooperatives for biodiesel,"Bioresource Technology, vol. 70, no. 1, pp. 81–87, 1999.

207. C. Manivannan and S. P. Sawan, "The supercritical state," in Supercritical Fluid Cleaning: Fundamentals, Technology, and Applications, pp. 1–2, William Andrew, 1998.

208. Z. Berk, Food Process Engineering and Technology, Elsevier, New York, NY, USA, 2009.

209. G. Brunner, "Supercritical fluids: technology and application to food processing," Journal of Food Engineering, vol. 67, no. 1-2, pp. 21–33, 2005.

210. H. Berg, M. Mågård, G. Johansson, and L. Mathiasson, "Development of a supercritical fluid extraction method for determination of lipid classes and total fat in meats and its comparison with conventional methods," Journal of Chromatography A, vol. 785, no. 1-2, pp. 345–352, 1997.

211. H. Saad and E. Gulari, "Diffusion of liquid hydrocarbons in supercritical $CO_2$ by photon correlation spectroscopy," Berichte der Bunsen-Gesellschaft für Physikalische Chemie, vol. 88, no. 9, pp. 834–837, 1984.

212. J. P. Friedrich and E. H. Pryde, "Supercritical $CO_2$ extraction of lipid-bearing materials and characterization of the products," Journal of the American Oil Chemists' Society, vol. 61, no. 2, pp. 223–228, 1984.

213. P. Tonthubthimthong, P. L. Douglas, S. Douglas, W. Luewisutthichat, W. Teppaitoon, and L. E. Pengsopa, "Extraction of nimbin from neem seeds using supercritical $CO_2$ and a supercritical $CO_2$-methanol mixture," Journal of Supercritical Fluids, vol. 30, no. 3, pp. 287–301, 2004.

214. W. E. Brewer, R. C. Galipo, K. W. Sellers, and S. L. Morgan, "Analysis of cocaine, benzoylecgonine, codeine, and morphine in hair by supercritical fluid extraction with carbon dioxide modified with methanol," Analytical Chemistry, vol. 73, no. 11, pp. 2371–2376, 2001

215. N. Aghel, Y. Yamini, A. Hadjiakhoondi, and S. M. Pourmortazavi, "Supercritical carbon dioxide extraction of Mentha pulegium L. essential oil," Talanta, vol. 62, no. 2, pp. 407–411, 2004.

216. R. M. Couto, P. C. Simões, A. Reis, T. L. Da Silva, V. H. Martins, and Y. Sánchez-Vicente, "Supercritical fluid extraction of lipids from the heterotrophic microalga Crypthecodinium cohnii,"Engineering in Life Sciences, vol. 10, no. 2, pp. 158–164, 2010.

217. R. L. Mendes, A. D. Reis, and A. F. Palavra, "Supercritical $CO_2$ extraction of -linolenic acid and other lipids from Arthrospira (Spirulina) maxima: comparison with organic solvent extraction," Food Chemistry, vol. 99, no. 1, pp. 57–63, 2006.

218. C. G. Laudani, M. Habulin, Ž. Knez, G. D. Porta, and E. Reverchon, "Lipase-catalyzed long chain fatty ester synthesis in dense carbon dioxide: kinetics and thermodynamics," Journal of Supercritical Fluids, vol. 41, no. 1, pp. 92–101, 2007.

219. D. Oliveira and J. V. Oliveira, "Enzymatic alcoholysis of palm kernel oil in n-hexane and $SCCO_2$," Journal of Supercritical Fluids, vol. 19, no. 2, pp. 141–148, 2001.

220. M. N. Varma and G. Madras, "Synthesis of biodiesel from castor oil and linseed oil in supercritical fluids," Industrial and Engineering Chemistry Research, vol. 46, no. 1, pp. 1–6, 2007.

221. M. N. Varma, P. A. Deshpande, and G. Madras, "Synthesis of biodiesel in supercritical alcohols and supercritical carbon dioxide," Fuel, vol. 89, no. 7, pp. 1641–1646, 2010.

222. A. Carrero, et al., "Hierarchical zeolites as catalysts for biodiesel production from Nannochloropsis microalga oil," Catalysis Today, vol. 167, no. 1, pp. 148–153, 2011.

223. E. A. Ehimen, Z. F. Sun, and C. G. Carrington, "Variables affecting the in situ transesterification of microalgae lipids," Fuel, vol. 89, no. 3, pp. 677–684, 2010.

224. B. D. Wahlen, R. M. Willis, and L. C. Seefeldt, "Biodiesel production by simultaneous extraction and conversion of total lipids from microalgae, cyanobacteria, and wild mixed-cultures," Bioresource Technology, vol. 102, no. 3, pp. 2724–2730, 2011.

225. W. Xiong, X. Li, J. Xiang, and Q. Wu, "High-density fermentation of microalga Chlorella protothecoides in bioreactor for microbio-diesel production," Applied Microbiology and Biotechnology, vol. 78, no. 1, pp. 29–36, 2008.

226. M. Lanza, W. L. Priamo, J. V. Oliveira, C. Dariva, and D. De Oliveira, "The effect of temperature, pressure, exposure time, and depressurization rate on lipase activity in $SCCO_2$," Applied Biochemistry and Biotechnology Part A, vol. 113, no. 1-3, pp. 181–187, 2004.

# Production And Use of Lipases in Bioenergy: A Review From the Feedstocks to Biodiesel Production

Bernardo Dias Ribeiro[1], Aline Machado de Castro[2], Maria Alice Zarur Coelho[1], and Denise Maria Guimarães Freire[3]

[1]School of Chemistry, Federal University of Rio de Janeiro, 21941-970 Rio de Janeiro, RJ, Brazil

[2]Biotechnology Division, Research and Development Center, 21941-915 Petrobras, Brazil

[3]Institute of Chemistry, Federal University of Rio de Janeiro, 21941-970 Rio de Janeiro, RJ, Brazil

## ABSTRACT

Lipases represent one of the most reported groups of enzymes for the production of biofuels. They are used for the processing of glycerides and fatty acids for biodiesel (fatty acid alkyl esters) production. This paper presents the main topics of the enzyme-based production of biodiesel, from the feedstock's to the production of enzymes and their application in esterification and trans esterification reactions. Growing technologies, such as the use of whole cells as catalysts, are addressed, and as concluding remarks, the advantages, concerns, and future prospects of enzymatic biodiesel are presented.

# LIPID FEED STOCKS

The main feed stocks which present paramount importance for the application of lipases are fats and oils. Such materials are primarily composed of triglycerides, which are glycerol esters with saturated and unsaturated fatty acids, from vegetable, animal, or microbial origins. One of the distinguishable characteristics between fats and oils is the occurrence of unsaturated and saturated fatty acids in the triglycerides: higher saturated fatty acids content (as examples in Figure 1), higher melting point, and the presence of remaining solids at room temperature are characteristics of a fat; on the other hand, oils usually present higher occurrence of unsaturated fatty acids, remaining in liquid state at room temperature. In addition to triglycerides, vegetable oils can present di- and monoglycerides, free fatty acids (FFAs), phosphatides, and unsaponifiable matter, such as carotenoids, phytosterols, tocochromanols, chlorophyll, triterpenic alcohols, and hydrocarbons [1–4].

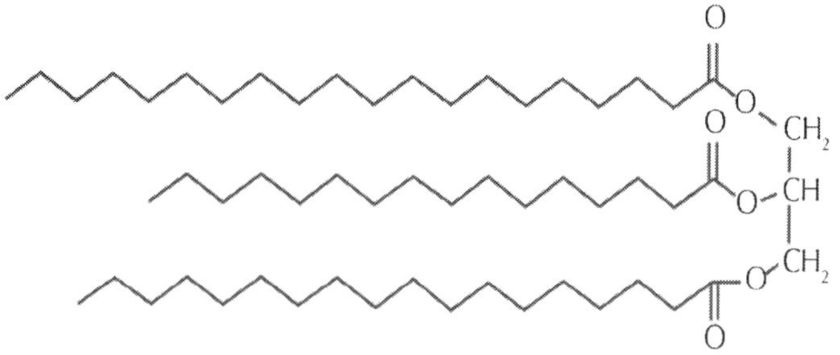

**Figure 1:** Schematic representation of a triglyceride with saturated fatty acids.

The role of fats and oils in plants is related to energy reserve, regarding their occurrence in seeds, and protection against water loss (by wax formation) and against mechanical injuries (by hormone generation), when such components appear in the leaves and fruits [2, 5].

Worldwide production of fats and oils was estimated in 174.6 million tons for the season 2010/2011. From that, 86% represent vegetable oils (Table 1), with soybean, palm, rapeseed, and sunflower seed as the major resources [6, http://lipidlibrary.aocs.org/, 2011]. In Brazil, some oilcrops, such as castor bean (Ricinus communis), jatropha (Jatropha curcas), crambe (Crambe abyssinica), macaw palm (Acrocomia aculeata), and oiticica (Licania rigida), have been explored as alternatives for biodiesel production due to their high tolerance to drought and frost, higher productivity on low-fertility soils, and great potential for the sustainable development of Brazilian Northeast [7, http://www.ruralbioenergia.com.br/, 2009]. Moreover, the use of raw materials with appropriated physicochemical characteristics and widely available enables cost reduction for the production of the biofuel, since the feedstock cost represents 70–88% of the final price of biodiesel [8].

**Table 1:** World oilcrops distribution [6]

| Fats and oils | World production (million tons) | Five major producers |
|---|---|---|
| Animal fat | 24.4 | USA, China, Brazil, Germany, and France |
| Coconut oil | 3.7 | Philippines, Indonesia, India, Vietnam, and Mexico |
| Cottonseed oil | 4.8 | China, India, Pakistan, Uzbekistan, and USA |
| Groundnut oil | 5.3 | China, India, Nigeria, Myanmar, and Sudan |
| Linseed oil | 0.6 | China, Belgium, USA, Ethiopia, and India |
| Maize oil | 2.3 | USA, China, Japan, Brazil, and South Africa |

| Olive oil | 2.9 | Spain, Italy, Greece, Syrian Arab Republic, and Tunisia |
|---|---|---|
| Palm kernel oil | 5.6 | Indonesia, Malaysia, Nigeria, Thailand, and Colombia |
| Palm oil | 23.9 | Malaysia, Nigeria, Thailand, Colombia, and Côte d'Ivoire |
| Rapeseed oil | 21.2 | China, Germany, India, Canada, and France |
| Safflower oil | 0.1 | India, USA, and Argentina |
| Sesame oil | 0.9 | Myanmar, China, India, Sudan, and Japan |
| Soybean oil | 36.0 | USA, China, Brazil, Argentine and India |
| Sunflower oil | 13.0 | Russian Federation, Ukraine, Argentine, Turkey, and France |

For the selection of a proper raw material for use as substrate for the production of biodiesel, some aspects should be observed, such as the following.

# Fatty Acids Profile

The fatty acids profile varies greatly between fats and oils and can be referred by distinct nomenclatures (Table 2). It can comprise from high concentration of saturated fatty acids, like in palm seeds, such as coconut (Cocos nucifera), palm kernel (Elaeis guineensis), and babassu (Orbignya oleifera) (Table 3), as well as animal fats (Table 4), to high content of monounsaturated fatty acids, commonly in oleaginous fruits (Table 5). Certainly, there are some exceptions of typical profiles, such as castor bean oil, which has a high content of ricinoleic acid; crambe, with high quantity of erucic acid; palm, with similar quantities of saturated and unsaturated fatty acids (Table 3) [9].

**Table 2:** Nomenclature of fatty acids [1, 9]

| Common name | Systematic name | Chemical structure[1] | Melting point (°C) |
|---|---|---|---|
| Lauric acid | Dodecanoic acid | 12:0 | 44.2 |
| Miristic acid | Tetradecanoic acid | 14:0 | 54.4 |
| Palmitic acid | Hexadecanoic acid | 16:0 | 62.9 |
| Palmitoleic acid | 9-Hexadecenoic acid | 16:1 | −0.1 |
| Stearic acid | Octadecanoic acid | 18:0 | 70.1 |
| Oleic acid | 9-Octadecenoic acid | 18:1 | 16.3 |
| Elaidic acid | 9-Octadecenoic acid | 18:1 | 43.7 |
| Vaccenic acid | 11-Octadecenoic acid | 18:1 | 44.0 |
| Linoleic acid | 9, 12-Octadecadienoic acid | 18:2 | −6.5 |
| -Linolenic acid | 6, 9, 12-Octadecatrienoic acid | 18:3 | −11.0 |
| -Linolenic acid | 9, 12, 15-Octadecatrienoic acid | 18:3 | −12.8 |
| Arachidic acid | Eicosanoic acid | 20:0 | 76.1 |
| Gadoleic acid | 9-Eicosenoic acid | 20:1 | 25.0 |
| Arachidonic acid | 5, 8, 11, 14-Eicosatetraenoic acid | 20:4 | −49.5 |
| Behenic acid | Docosanoic acid | 22:0 | 80.0 |
| Erucic acid | 13-Docosenoic acid | 22:1 | 33.4 |

[1]x: y nomenclature, where x represents the total number of carbon atoms and y represents the number of unsaturated bonds.

**Table 3:** Fatty acids profile of oilcrops [1, 9]

| Fatty acids[1] | Palm kernel | Soybean | Jatropha curcas | Crambe | Rapesed | Sunflower | Castor bean | Babassu |
|---|---|---|---|---|---|---|---|---|
| 12:0 | 41–55 | NR | NR | NR | NR | NR | NR | 40–55 |
| 14:0 | 14–18 | NR | NR | NR | <0.2 | <0.5 | NR | 11–27 |
| 16:0 | 6.5–10.0 | 7–14 | 10–17 | 1.8–2.0 | 2.5–6.5 | 3.0–10.0 | 1.1 | 5.2–11.0 |
| 16:1 | NR | <0.5 | NR | NR | <0.6 | <0.1 | 0.2 | NR |
| 18:0 | 1.3–3.0 | 1.4–5.5 | 5–10 | 0.7–1.0 | 0.8–3 | 1–10 | 1 | 1.8–7.4 |
| 18:1 | 12–19 | 19–30 | 36–64 | 16.0–17.2 | 53–70 | 14–35 | 3.3[2] | 9–20 |
| 18:2 | 1–3.5 | 44–62 | 18–45 | 8.0–8.7 | 15–30 | 55–75 | 3.6 | 1.4–6.6 |
| 18:3 | NR | 4–11 | NR | 5.2–7 | 5–13 | <0.3 | 0.32 | NR |
| 20:0 | NR | <1.0 | NR | 3.4 | 0.1–1.2 | <1.5 | 0.4 | NR |
| 20:1 | NR | <1.0 | NR | NR | 0.1–4.3 | <0.5 | NR | NR |
| 22:0 | NR | <0.5 | NR | NR | <0.6 | <1.0 | NR | NR |
| 22:1 | NR | NR | NR | 56–66 | 0.7 | NR | NR | NR |
| % oil | 45–50 | 18–20 | 26–35 | 35–60 | 40–50 | 22–36 | 35–55 | 65–68 |

[1] x: y nomenclature, where x represents the total number of carbon atoms and y represents the number of unsaturated bonds; [2] 80–90% ricin oleic acid (similar to oleic acid plus a hydroxyl group in position 12R). NR: not reported.

**Table 4:** Fatty acid profile from animal origin [9]

| Fatty acids | Butter | Lard | Tallow |
|---|---|---|---|
| <14:0 | 11.0–23.8 | 0.5 | 0.9 |
| 14:0 | 8.2–12.0 | 1.3 | 3.0–3.7 |
| 16:0 | 21.3–29.0 | 23.8–25.0 | 24.9–27.0 |
| 18:0 | 9.8–13.0 | 12.0–13.5 | 7.0–18.9 |
| 16:1 | 1.8–2.0 | 2.7–3.0 | 4.2–11.0 |
| 18:1 | 20.4–28.0 | 41.2–45.0 | 36.0–48.0 |
| 18:2 | 1.8 | 10.0–10.2 | 3.1 |
| 18:3 | 1.2 | 1.0 | 0.6 |
| % fat | 2–5 | 70–95 | 70–95 |

**Table 5:** Fatty acids profile of oleaginous fruits

| Fatty acids[1] | Buriti | Olive | Avocado | Palm |
|---|---|---|---|---|
| 12:0 | NR | NR | NR | 0.1–1.0 |
| 14:0 | 0.1 | 0.7 | <0.13 | 0.9–1.5 |
| 16:0 | 17.3–19.3 | 10–11.7 | 19.8–22.7 | 41.8–46.8 |
| 18:0 | 1.9–2.0 | 2.1 | 0.5–1.0 | 4.2–5.1 |
| 20:0 | NR | 0.48 | NR | 0.2–0.7 |
| 16:1 | NR | 1.45 | 3.9–5.6 | 0.1–0.3 |
| 18:1 | 73.3–78.7 | 73.8–78 | 60–71 | 37.3–40.8 |
| 18:2 | 2.4–3.9 | 7.0–9.8 | 7.1–15.3 | 9.1–11.0 |
| 18:3 | 2.2 | NR | 0.4–1.0 | <0.6 |
| % oil | 8–18 | 15–40 | 4–25 | 20–24 |
| References | [10] | [1] | [11] | [12] |

Additionally to the vegetal and animal sources of lipids and fats presented in Tables 3–5, fatty acids can also come from microbial origin. As recently revised by Li et al. [21], yeasts from Cryptococcus, Lipomyces,Rhodosporidium, Rhodotorula, Trichosporon, and Yarrowia genera, as well as filamentous fungi and bacteria, can reach 53% of lipids content in its dry mass, with evidence for major appearance of palmitic and oleic acids. In another work, the profiles of fatty acids of the microalgaes Spirulina sp., Scenedesmus obliquus,Chlorella vulgaris, and C. kessleri were determined [22]. The authors observed the prevalence of saturated fatty acids (lauric, miristic, palmitic, and stearic acids), with contributions of up to 46% for the total fatty acids content.

These diversified profiles of fatty acids from different origins contribute for the generation of biofuels with different properties. For example, the higher the size of fatty acid hydrocarbon chain, the higher the cloud point and the cold filter plugging point. Therefore, due to the necessity of heating before ignition, it becomes difficult the use of a biodiesel with such characteristic in regions with low environment temperature.

Another factor concerning the use of unsaturated fatty acids for the production of biodiesel is that the fewer the double bonds in the molecules, the higher the cetane number of the biofuels (which, in turn, means a better quality of their combustion). Moreover, larger quantities of unsaturated bonds turn molecules more chemically unstable. This can cause some inconvenience due to the biofuel oxidation, degradation, and polymerization (resulting in low cetane number or formation of solid residues), if improperly stocked or transported. Then, in general, a biodiesel with high quantities of esters derived from monounsaturated fatty acids (e.g., oleic or ricin oleic acids) presents better results as a fuel [23].

## Fats and Oils Processing

Animal fat processing is named rendering, where carcasses with fatty material are heated with hot water or steam to release fats, with subsequent separation by centrifugation or by surface removal. The vegetable oil processing is comprised of some steps, including mechanical pretreatment (cleaning, sorting, and comminution), heating, dehydration, mechanical pressing and/or solvent extraction, miscella distillation, meal desolventization, and refining [1, 9].

For biodiesel production, the oil refining processes play an important role in the yield of the conversion steps, since oil impurities, such as water, phosphatides, and pigments, can affect the conversion of triglycerides to esters due to excessive emulsification of the reaction mixture and difficulties in biodiesel separation, amongst others [1, 9].

Another important factor during feedstock processing is the valorization of coproducts. Such approach can contribute to the profits of an industrial plant, thus bettering the viability of biodiesel. As a classical example, soybean meal generated during soybean oil extraction is already used for protein and is flavones extraction [24], and its main phospholipid, lecithin, separated in the degumming step, is used as natural emulsifier [25].

# ENZYME PRODUCTION AND CHARACTERISTICS

Lipases are enzymes classified as hydrolases (glycerol ester hydrolase, E.C. 3.1.1.3) and act on ester bonds of several compounds, with acylglycerols being the most proper substrates, catalyzing reactions of hydrolysis, synthesis, and trans- and interesterification (Figure 2). Lipases are more active in insoluble substrates, especially triglycerides made of long-chain fatty acids with over 10 carbon atoms, while esterases are active in soluble substrates, especially simple esters, such as ethyl acetate and triglycerides made of short-chain fatty acids with less than six carbon atoms. Esterases follow Michaelis-Menten kinetics, while lipases need a minimum substrate concentration to show high activity levels [26].

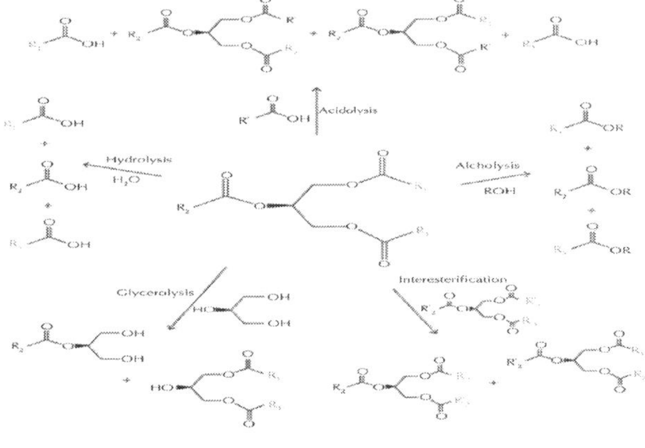

**Figure 2:** Reactions catalyzed by lipases.

Due to the similarity of the catalytic triad found in lipases compared to those observed in serine proteases, the most widely accepted hypothesis is that the mechanism of lipase catalysis is similar to that of serine protease catalysis [27]. It is believed that the kinetic mechanism of lipases does not depend on the type of reaction being catalyzed (hydrolysis, acidolysis, transesterification, etc.).

The reaction begins with a nucleophilic attack on the carbon from the ester bond of the susceptible substrate by hydroxyl group in the serine residue of the active site, forming an acyl-enzyme complex and releasing alcohol from the lipid. Later, the acyl-enzyme complex is hydrolyzed, releasing the lipase regenerated. Figure3 shows the stages of the reaction catalyzed by the lipase and its intermediates.

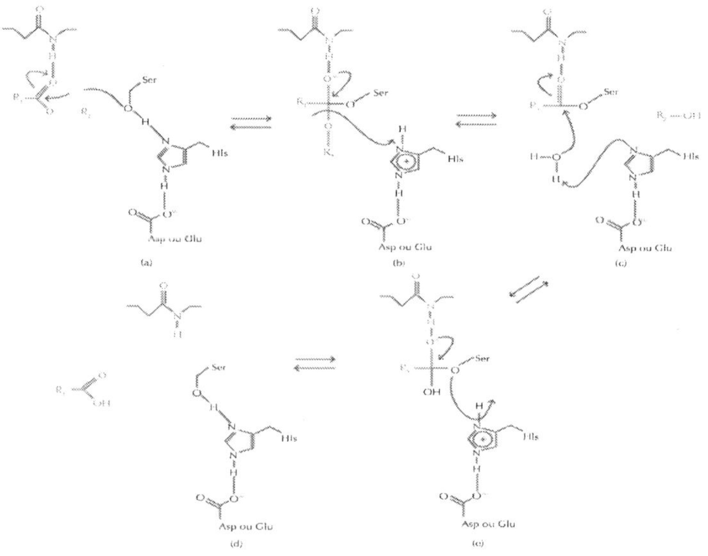

**Figure 3:** Mechanism of the hydrolysis reaction of ester bonds catalyzed by esterases and lipases. The catalytic triad and water are shown in black; the oxyanion whole residues are in blue; the substrate is in red. (a) Nucleophilic attack of the serine hydroxyl on the carbonyl carbon of the susceptible ester bond; (b) tetrahedral intermediate; (c) acyl-enzyme intermediate and nucleophilic attack by water; (d) tetrahedral intermediate; (e) free enzyme [90].

Furthermore, characteristics such as stability in the presence of organic solvents, no necessity of cofactors for their action and high enantioselectivity, turn lipases into a group of enzymes with one of the major technological interests [28–30].

Lipases occur widely in nature and can be produced by many microorganisms and higher eukaryotes. In animals, lipases obtained from pig and human pancreas are best known and more investigated than all other lipases. In these organisms, they are engaged in several lipid metabolism steps, including fat digestion, adsorption, reconstitution, and in lipoproteins metabolism. In plants, lipases are present in higher plants seeds, as castor bean and canola (Brassica napus). They are also found in several plants' energy reserve tissues [28, 31–33]. However, for the production of industrial enzymes, microorganisms are the preferred source, once they have shortest generation time, high yield of conversion of substrate into product, great versatility to environmental conditions and, simplicity in genetic manipulation and in cultivation conditions. Due to habitats' multiplicity, microorganisms usually produce various lipases types, with distinct specificity regarding to substrate utilization and also to optimum pH and temperature range. Lipases can be produced by bacteria, filamentous fungi, and yeasts, allowing these microorganisms to use lipids from animal or vegetable origin as carbon and energy sources for their growth. Though many microorganisms have been reported in literature as lipase producers, the genera Candida, Rhizopus, and Pseudomonas are considered the main industrial sources of lipases. The yeast Candida rugosa is the most employed microorganism for lipase production [30]. Table 6 gives an overview on recent literature regarding lipases production.

**Table 6:** Insight into recent literature on microbial lipase production

| Microorganism | Raw material | Type of fermentation | Maximum activity (time of fermentation) | Reference |
|---|---|---|---|---|
| A. niger 11T53A14 | Wheat bran | SSF | 62.7 $Ug^{-1}$ (48 h) | [13] |
| Penicillium sp. | Olive oil | SmF | 21.0 $UmL^{-1}$ (120 h) | [14] |
| Rhizopus oryzae NRRL 3562 | Coconut oil | SSF | 96.2 $Ug^{-1}$ (115 h) | [15] |
| Bacillus subtilis OCR-4 | Ground nut oil cake | SSF | 4.5 $Ug^{-1}$ (48 h) | [16] |

| Burkholderia cepaciaLTEB11 | Sugarcane bagasse and sunflower seed meal | SSF | 234 Ug$^{-1}$ (96 h) | [17] |
|---|---|---|---|---|
| Rhizopus chinensis | Wheat bran, wheat flour, and olive oil | SSF | 24.4 Ug$^{-1}$ (72 h) | [18] |
| Pseudozyma hubeiensisHB85A | Soybean oil | SmF | 5.3 UmL$^{-1}$ (18 h) | [19] |
| P. chrysogenum | Grease waste and wheat bran | SSF | 46 UmL$^{-1}$ (168 h) | [20] |

SSF: Solid-state fermentation; SmF: Submerged fermentation.

The use of lipases in industry is still limited by the cost of commercial enzymes, especially when large quantities of enzyme are required and when the final product is of low added value. There is therefore a considerable interest in reducing the cost of producing these biocatalysts. The use of solid-state fermentation (SSF) as a production system is one way of reducing enzyme production costs, especially because agroindustrial waste can be used as a culture medium.

A comparative economic analysis showed that the production of lipase from Penicillium restrictum by SSF is more economically feasible than its production by submerged fermentation (SmF), with a production cost for the former being found to be 68% lower and a payback time of 1.5 years [54].

Other advantages of producing enzymes by SSF have been highlighted alongside the reduced production costs. In studies of lipase production by the fungus Penicillium restrictum using SSF and SmF, different significant physiologies were observed between the two systems when simple (oleic acid and glucose) and complex (olive oil and starch) sources of carbon were used, with a reduction in catabolite repression being observed for SSF [55].

Lipases from different microorganisms have been produced using SSF with different solid wastes, such as lipase from Penicillium restrictum in babassu cake [55, 56]; lipase from P. simplicissimum in bagasse cake, soybean cake, and castor bean cake [57–61]; lipase from Candida rugosa in rice flour [62]; lipase fromRhizopus homothallicus in sugarcane bagasse [63, 64]; lipase from Aspergillus niger in wheat bran and sesame seed cake [65, 66]; lipase from Rhizopus rhizopodiformis and Rhizomucor pusillus in olive oil cake and sugarcane bagasse [67]; lipase from Rhizopus oligosporus in a variety of cakes [68]. These lipases were produced by SSF on a bench scale, mostly using tray bioreactors, and yielded high productivity rates.

There are no pre-established procedures in the literature for predicting the performance and design of SSF bioreactors. For this reason, large-scale systems have generally been developed from the results obtained from bench-scale or pilot systems. Ideally, a large-scale system should operate in the same way and with the same performance as a bench-scale system although this is often not the case for SSF processes [69]. The main limiting factor on scaling up such processes is heat transfer, which depends on the stage of fermentation, and the design and operation mode of the bioreactor [70–72]. Some mathematical models have been developed to describe the growth kinetics of the microorganisms under different operating conditions and to describe heat and mass transfer in tray bioreactors [73], fixed-bed bioreactors [70, 74], rotating drum bioreactors [75, 76], shaking reactors [77], and fluidized bed reactors. These models could be used as inputs for designing the scale-up of such systems.

In addition to the reduction of the costs related to fermentation step for industrial-scale production of lipases, the strategies to recover and purify lipases must also be as low as possible and should also be rapid, give high yields, and ideally be easy to scale up.

Lipases are enzymes that are known to be strongly hydrophobic, because of the presence of alkyl groups on the surface of their structure [30]. Generally, a good first step for lipase purification is the use of hydrophobic-interaction chromatography. Normally, prepurification involves precipitation with ammonium sulphate, and ion-exchange chromatography and gel filtration are also widely used [78–80].

Rua et al. [79] studied the production and purification of a thermostable alkaline lipase from Bacillus thermocatenulatus in Escherichia coli. The purification stages were done in butyl sepharose (hydrophobic bed) and TSK G3000 (gel filtration), giving a purification factor of 125 and a yield of 32%.

A lipase from Aspergillus niger F044 was purified by precipitation with ammonium sulphate, DEAE-Sepharose FF (ion exchange), and

Sephadex G-75 (gel filtration). A yield of 33% was obtained, while the purification factor was 73 [37].

A lipase from Penicillium simplicissimum produced by submerged fermentation was purified in a five-step process [81]. First, the culture was concentrated using a 10 kDa membrane, then it was precipitated with ammonium sulphate. After concentration and prepurification, the sample was injected in sequential chromatography steps on phenyl sepharose CL-4B (hydrophobic interaction), Ultrogel AcA-54 (gel filtration), and hydroxyapatite (ion exchange). The resulting purification factor was 788, and the yield was 20%.

In order to purify a lipase from Penicillium camembertii U-159, Isobe et al. [82] used ethanol precipitation and ammonium sulphate precipitation as the first and second steps. A sequence of chromatography steps followed, using DEAE-sepharose (ion exchange), amino octyl sepharose (hydrophobic interaction), hydroxyapatite (ion exchange), and concanavalin-A sepharose (affinity). The yield obtained was 27%, and the purification factor was 213.

Cunha et al. [83] studied the purification/immobilization of a "pool" of lipases from P. simplicissimum produced by SSF using babassu cake as a culture medium. The process undertaken by means of sequential immobilization in hydrophobic supports (butyl, phenyl, and octyl agarose) resulted in three fractions with distinct thermal stability, specificity, and enantioselectivity properties.

Depending on the source, lipases can present molar mass ranging from 20 to 75 kDa, enzymatic activity at pH between 4 and 9 and at temperatures since 27 until 70°C. Lipases are usually stable in neutral aqueous solutions at room temperature, presenting, in most cases, an optimal activity at 30–40°C. However, its thermostability varies considerably depending on the origin, and, according to Castro et al. [29], microbial lipases present the best thermo stability.

Most commercial lipolytic preparations are composed by a mixture of various isozymes, in different proportions, such as those obtained from Candida rugosa, Pseudozyma (formerly Candida) antarctica, Rhizopus niveus, and Chromobacterium viscosum, among others. Each isoform has different properties (e.g., molar mass, specificity, stereoselectivity, glycosylation extension, and substrate preference) [28, 84, and 85]. The main sources of lipases and their properties are described in Table 7.

**Table 7:** Sources of lipases and optimal conditions for their action

| Sources | pH | T (°C) | Reference |
|---|---|---|---|
| Candida rugosa | 5–8 | 35–50 | [34] |
| Pseudozyma antarctica A | 6–10 | 35–70 | [35] |
| Thermomyces lanuginosus | 6–9 | 30–50 | [36] |
| Aspergillus niger | 6–8 | 40–55 | [37] |
| Pseudomonas aeruginosa | 5.5–7.5 | 35–45 | [38] |
| Bacillus subtilis | 8–10 | 30–40 | [39] |
| Geotrichum candidum | 6.5–8.0 | 32–42 | [40, 41] |
| Streptomyces rimosus | 8.5–10.0 | 45–60 | [42] |
| Yarrowia lipolytica | 4–7 | 30–45 | [43–45] |
| Rhizopus niveus | 5–7 | 30–45 | [46] |
| Rhizomucor miehei | 6.5–7.5 | 30–40 | [47] |
| Porcine pancreatin | 6–9 | 40–55 | [48] |
| Castor bean (Ricinus communis) | 4.0–4.5 | 30–35 | [49, 50 |

For industrial applications, the specificity of lipase is a crucial factor. This enzyme can present specificity regarding the substrate (fatty acid or alcohol), including the differentiation of isomers. Lipases can be divided into three groups based on their specificity.

1.  Nonspecific lipases (such as those produced by Candida rugosa, Staphylococcus aureus,Chromobacterium viscosum, Thermomyces lanuginosus, and Pseudomonas sp.): They cleave acylglycerol molecules randomly generating FFAs and glycerol, as well as mono- and diglycerides as intermediates. In this case, the products are similar to those produced by chemical catalysis, but with less thermodegradation, due to the lower temperature used for the reaction, when compared to chemical processes [29, 46, 86].

2.  1, 3-specific lipases (e.g., from Aspergillus niger, Mucor javanicus, Rhizopus delemar, Rhizopus oryzae, Yarrowia lipolytica, Rhizopus niveus, and Penicillium roquefortii): They release fatty acids from positions 1 and 3 of a glyceride and from, for this reason, products with different compositions of those obtained by nonregioselective lipases, or even by chemical catalysts. Generally, the hydrolysis of triglycerides to diglycerides is much faster than those into monoglycerides [29, 46, and 87].

3.  Fatty acid-specific lipases: they act specifically on the hydrolysis of esters, which have long-chain fatty acids with double bonds in cis position on carbon 9. Esters with unsaturated fatty acids, or without double bond in carbon 9, are slowly hydrolyzed. This type of specificity is not common among the lipases and probably the most studied example of this case is the lipase from Geotrichum candidum [29, 46, 87–89].

The study of substrate specificity is also of great importance for the application of lipases in biodiesel production, since it is a valuable input for the selection of the proper enzyme based on the composition of the raw material. Gutarra et al. [58] evaluated the substrate specificity of an acidic lipase produced byPenicillium simplicissimum, observing the highest activities on tricaprin (C8:0) and tricapryçin (C10:0), which were 83 and 92% higher, respectively, than those detected in the model substrate (olive oil).

Lipases can also be stereospecific, where one of the isomers of a racemate is hydrolyzed preferentially over another, or even the formation of one isomer can be catalyzed selectively from prochiral precursors such asmeso-diester or meso-diol compounds. Some examples are lipases from Burkholderia cepacia, Pseudozyma antarctica, Candida rugosa, and Rhizopus delemar [88, 89].

# ENZYME-CATALYZED PROCESSES FOR THE PRODUCTION OF BIODIESEL

The main technology for biodiesel production in Brazil and in the world is homogenous alkaline transesterification (or alcoholysis). In this reaction, an alcohol (usually methanol or ethanol), with a molar basis, is added to the oil or fat and, in the presence of a catalyst (Brønsted acids or bases), a mixture of glycerin and alkyl esters of fatty acids is generated, which is called biodiesel (Figure 4). However, alkaline catalysts, especially sodium hydroxide, became dominant for the production of biodiesel, due to their lower costs and faster kinetics [9, 23, 91, and 92].

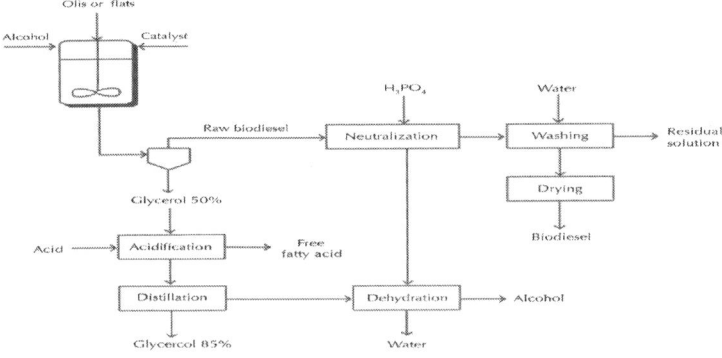

(a)

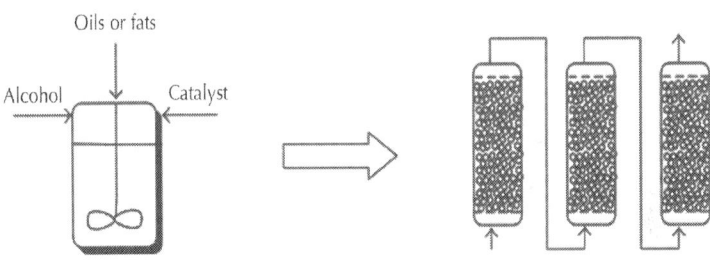

**Figure 4:** Simplified flowsheet for the production of biodiesel. (a) Classical industrial chemical process; (b) alteration in reactor design for biocatalysis.

However, homogenous alkaline trans esterification presents some disadvantages over enzyme-catalyzed processes, such as the need of raw materials (refined oils and alcohols) virtually free of fatty acids, phosphatides, and water; excess of alcohol and catalyst to avoid reversible reactions, which in turn makes difficult the separation of biodiesel and glycerin. Therefore, alternative catalysts have been studied, such as organic bases, metallic complexes, oxides, aluminosilicates, and enzymes. Their main characteristics are that they are easily recycled and the absence of soap formation, which facilitates the products separation at the end of alcoholics [91–93].

The use of biocatalysts has, therefore, advantages over chemical processes, and these include esterification of both triglycerides and fatty acids; generation of a cleaner glycerol; reuse, mostly in the case of an immobilized lipase utilization. However, some problems still need to be resolved, as high cost of lipases and possible inhibition in the presence of short-chain alcohols, glycerol, and other impurities in the raw material [23, 91, 93, and 94]. In the case of biocatalysts, the schematic flow sheet of Figure 4(a) can also be applied, but it can also be necessary to use immobilized enzymes, for the reasons shown above. Due to kinetic disadvantages, it can be necessary also to use more sequential reactors in order to achieve the residence time of the feedstock in the presence of the enzymes, for a desired conversion (Figure 4(b)). For biodiesel production by enzymatic catalysis, some factors should be considered and some topics should be covered, which can be divided into aspects for current and prospective approaches. These aspects are detailed as follows.

# Current Aspects

## *Refining Step*

In enzymatic Tran's esterification, higher yields are achieved for biodiesel production when refined oil is used, compared to crude oils. This is due to the presence of phospholipids in the no refined oil, which affect the interaction between lipase and substrate, since they possibly occlude the pores of the support, in the case of using an immobilized enzyme. Therefore, at least the oil degumming step should be conducted before Tran's esterification reactions, in order to obtain a better production of biodiesel [23, 91].

The oil degumming is traditionally done using chemical and physical processes, such as water degumming, ultrafiltration, and mainly acid (phosphorous or citric) treatment [95, 96]. However, since the 1990s, enzyme-catalyzed degumming has been reported as a potential alternative to the conventional processes, and this comprises the use of phospholipases, which are classified into four groups [97]. Phospholipases types A1 (E.C. 3.1.1.32) and A2 (E.C. 3.1.1.4) catalyze the cleavage of ester bonds in phospholipids, thus releasing FFAs and contributing for the increase of the overall yield of biodiesel. There are also phospholipases types C (E.C. 3.1.4.3) and D (E.C. 3.1.4.4), but these are involved in the breakdown of phosphate bonds in phospholipids and do not contribute to the increase in FFAs content of the oil [98]. Enzymatic degumming is done at mild temperature (40–45°C) and pH of about 4.5–5.0, for a period of 2–4 h [97, 99].

## Free and Immobilized Enzymes

The use of free enzymes for biodiesel production results in technical limitations, and it is practically unreliable, due to impossibility of their recovery and reuse, which in turn increases the production costs of the process, besides promoting the product contamination with residual enzyme. These difficulties can be overcome by the use of these enzymes in its immobilized form, allowing the reuse of biocatalyst several times, reducing costs, and further improving the quality of the product [100]. There are several techniques cited for lipases immobilization, such as adsorption, covalent bonding, entrapment, encapsulation, and cross-linking, but they will not be discussed in details in this paper, since there is a recently published review focusing on this issue [100]. In this context, Nielsen et al. [101] revised technical and economic aspects of biodiesel production, concluding that for enzyme-catalyzed biodiesel viability, using immobilized lipases, the enzymes must be stable for the production of 1200–7400 kg of biodiesel for each kg of enzyme preparation, depending on the substrate source and lipase used.

The confinement or physical location of an enzyme in a given region of a defined space, while maintaining their catalytic activity, which can be used repeatedly and continuously due to the ease of its separation from the reaction medium, comprises its immobilization [102].

The catalytic activities of enzymes and other features may change depending on the type of the immobilization technique (adsorption, covalent bound, entrapment, encapsulation, and cross-linking) and the interaction strength between enzyme and carrier used which may, in some cases, cause structural distortions in the protein. Still, the catalytic activity of the enzyme in a particular medium can be changed by increasing or decreasing stirring due to the support fragmentation by interaction between agitation system and support [100].

Thus, it is possible to occur some activity loss during transesterification reaction, even when immobilized lipases are used, and this is more probably due to enzyme leaching than to enzyme inactivation. On the other hand, if such leaching does not occur and the enzyme remains bound to the support, the increase of contact surface may help in raising mass transfer, thereby increasing the efficiency of the enzyme as a catalyst. Lipases from different sources have been immobilized and used in biodiesel production, but the most commonly reported were obtained from Pseudozyma antarctica and Thermomyces lanuginosus [8,103].

The use of fixed bed reactors with immobilized lipases is a more suitable solution for continuous production of biodiesel, since the enzymes suffer lower shear stress compared with the batch process. Pandey [91] reported the methanolysis of a waste oil mixture (containing 1980 ppm of water and 2.5% FFAs), using immobilized lipase from C. antarctica, and considering 3 steps of substrate addition. The highest biodiesel yield observed was 90.4%. In another example, the same author reported the ethanolysis of a fat in a recirculating packed bed reactor (flow of $1.8 Lh^{-1}$) using a phyllosilicate sol-gel-immobilized lipase fromBurkholderia cepacia, and in this case, a yield of 96% was observed. Over the subsequent four cycles evaluated, the yield was maintained in at least 90%.

## *Use of Solvents*

The enzymatic production of biodiesel can be performed using organic solvents (usually hexane, heptanes, or petroleum ether) or simply using the mixture of substrates (lipids and alcohol) depending on the size of the chain of alcohol. If methanol or ethanol is used, a solvent can facilitate the oil solubility in alcohol and also decrease the viscosity of the relational mixture, but there will be an additional cost for its removal (distillation or extraction) after the reaction [8, 94, 102].

Knothe et al. [23] reported the biodiesel production from sunflower oil using petroleum ether as solvent with immobilized lipase from Pseudomonas fluorescents, reaching 99% yield when the alcohol used was methanol, ethanol, or 1-butanol. In the absence of solvent, yields were reduced to 3%, 70, and 76%, respectively.

Soetaert and Vendome [104] reported the use of the lipases from Mucor miehei and C. antarctica in the transesterification of various oils using hexane as solvent and found that the lipase from M. miehei is more efficient in converting primary alcohols (methanol, ethanol, propanol, and 1-butanol) with yields between 95 and 98%, whereas lipase from C. antarctica is more proper for the conversion of secondary alcohols (isopropanol and 2-butanol) with yields between 61 and 84%. In the absence of the solvent, the yields of methyl and ethyl esters decreased, particularly when methanol was used, with yield reduction up to 19%.

The use of solvents with intermediate polarity (such as t-butanol, 1, 4-dioxane) has been suggested to achieve a better dissolution of the alcohol for transesterification (particularly methanol, due to its higher inhibitory effect over lipases) and oil, without affecting the lipase activity [102]. Ranganathan et al. [105] reported the use of t-butanol as a solvent for the transesterification of cottonseed oil using the commercial preparation Novozym 435 for 24 hours at 55°C and achieved a yield of 97%, maintaining the lipase activity by 95% of the initial activity during 500 h of continuous operation.

## Type of Alcohol and Adsorption Agent

Many types of alcohol can be used in the reaction of enzymatic transesterification. Some examples are shown in Table 8.

Table 8: Reported conditions for enzymatic Trans esterification of some fats and oils [51–53]

| Alcohol | Lipase source | Feedstock | Solvent | Yield (%) |
|---------|---------------|-----------|---------|-----------|
| Methanol | C. antarctica | Rapeseed oil | Hexane | 98 |
| Methanol | C. antarctica | Cottonseed oil | — | 92 |
| Methanol | C. antarctica | Cottonseed oil | t-Butanol | 97 |
| Methanol | C. antarctica | Degummed Soybean oil | — | 94 |
| Methanol | T. lanuginosus | Soybean oil | — | 90 |
| Ethanol | P. cepacia | Tallow fat | — | 95 |
| Propanol | P. fluorescens | Sunflower oil | 1,4-Dioxane | >95 |
| 2-Ethyl-1-hexanol | C. rugosa | Rapeseed oil | — | 97 |
| Methanol | P. cepacia | Palm kernel oil | — | 15 |
| Ethanol | P. cepacia | Palm kernel oil | — | 72 |
| Methanol | Mucor miehei | Soybean oil | Hexane | 75 |
| Ethanol | M. miehei | Soybean oil | Hexane | 97 |

| Methanol | M. miehei | Tallow fat | Hexane | 95 |
|---|---|---|---|---|
| Ethanol (96%) | M. miehei | Tallow fat | Hexane | 98 |
| Anhydrous ethanol | M. miehei | Tallow fat | Hexane | 68 |
| Propanol | M. miehei | Tallow fat | Hexane | 24 |
| Butanol | M. miehei | Tallow fat | Hexane | 20 |
| Propanol | C. antarctica | Tallow fat | Hexane | 61 |
| Butanol | C. antarctica | Tallow fat | Hexane | 84 |

The molar ratio between substrates is a variable with large influence on the biodiesel synthesis. Excess of alcohol, in relation to the stoichiometric ratio of 3/1, is used to ensure higher reaction rate and to minimize diffusional limitations. However, excessive levels of alcohol, mainly those with short chains, may inhibit the enzyme by increasing the polarity of the medium, which reduces the stabilization and removes the water layer associated with the immobilized enzymes [94]. This effect was observed by Brusamarelo et al. [106], for the transesterification of soybean oil. The optimal ethanol to oil ratio was 6/1, where 93% conversion was achieved after 480 min of reaction. When increased ethanol to oil ratios (9/1 and 12/1) were used, the conversion, considering the same time of catalysis, dropped to 51 and 55%, respectively.

Hence, the gradual addition of alcohol can maintain lipase stable for a longer period. Pandey [91] reported that the gradual addition of methanol (in 3 steps, every 16 h of reaction), maintaining the same enzyme in the bioreactor, resulted in a yield of 98% of biodiesel, and that the conversion was kept over 95% during 50 cycles. Another aspect that may accelerate the methanolysis is the preincubation of lipase in ester or oil [104].

The addition of silica to the reaction medium provides a positive effect on yield, due to the absorption of glycerol and water, thus reducing lipase inhibition [23]. One example cited by Robles-Medina et al. [102] is the use of 6% (w/w) of silica-gel in the reaction mixture along with the commercial preparation Lipozyme TL. The use of silica increased the yield from 66 to 90%. Another possibility would be to remove the glycerol by dialysis or its dissolution in isopropanol or t-butanol [102, 105].

# Alternative Donors of Acyl Group

As the use of methanol and ethanol can promote lipase inhibition, the use of alternative donors of acyl group, such as methyl acetate, ethyl acetate, and propan-2-ol, is being studied, since their use avoids the production of glycerol as a by-product, which blocks the porous support and lipase active sites. Novozym 435 was tested for biodiesel production using several oils and the donors of acyl groups cited above, and the results observed were yields always above 90% [53, 91].

# Water Addition

Alcoholic's reactions do not involve water as a reagent. However, the control of water content in the reaction system is important for some reasons: lipase requires a minimum amount of water to maintain its active conformation; an excess of water may promote the hydrolysis of the substrate and generate diffusion limitations of substrate, thus reducing the biodiesel yield; the water can influence negatively the reaction when methanol or ethanol is used but does not affect the reaction when higher-chain alcohols are considered [93, 94, 102].

Draco et al. [93] reported the use of lipase from Pseudomonas cepacia immobilized on polymeric sol-gel matrix in the transesterification of tallow oil at 40°C for 1 h using a mixture of 10 g of fat, 3 g of lipolytic preparation, 3 g of methanol or 5 g ethanol, and different amounts of water. The authors observed that when water concentrations below 0.2% were used, the conversion was significantly decreased and, after the reuse of lipase during 11 cycles, the activity was decreased by 10%.

Pandey [91] reported the use of lipase from Chromobacterium viscosum in the transesterification of jatropha oil with a 10% enzyme dosage. When the biocatalyst in a free form was used, it was observed a yield of 62%, whereas when the enzyme was immobilized on Celite 545, 71% of yield was achieved. By adding 1% of water to the free enzymatic preparation and 0.5% of water to the immobilized enzyme, the biodiesel yields raised to 73% and 92%, respectively.

## Temperature

Generally, the higher the temperature, the higher the reaction rate of alcoholysis or transesterification, until reaching the temperature of inactivation of lipases (usually above 60°C). This approach is valid mainly for systems in which the enzyme is used just once or few times. When enzyme reuse is considered, high temperatures, which can be suitable for short-term use of the enzymes, may be not the most proper, since the half-life time of lipases decreases with increases in temperature.

Matassoli et al. [107] investigated the influence of temperature in ethanolysis of crude palm oil catalyzed by Lipozyme TL IM (3% w/w) with a molar ratio ethanol/oil of 3/1 and gradual addition of ethanol, observing the best result at 50°C. For the evaluation of the effect of temperature on lipase-catalyzed biodiesel production, a semiempirical model was proposed by Brusamarelo et al. [106]. The authors investigated soybean biodiesel production using the commercial product Novozym 435, within the range of 45–70°C, observing the highest yield (92%) when 65°C was used.

## Enzyme Type and Dosage

The amount of enzyme added to reaction is also an important factor for biodiesel production, because it affects reaction rate (typically, the higher the enzyme dosage, the faster the reaction), but there is a limit in which the addition of enzyme does not alter anymore the rate of product formation or that the amount turns the process more economically prohibitive. In this context, enzyme-catalyzed biodiesel production was investigated using dosages of C. antarctica lipase B (Novozym 435) from 1 to 20% (w/w) [106]. The authors observed the highest conversion of triglycerides to ethyl esters (93%) when 10% (w/w) of the immobilized enzyme was used.

Regarding the effect of lipase specificity, Pandey [91] reported the use of some specific and non-specific lipases (from C. rugosa, P. cepacia, and P. fluorescens) in biodiesel production. The non-specific lipases promoted the highest yields of methyl esters when a molar ratio of 3/1 of methanol/oil was used. Specific lipases need gradual addition of methanol to achieve high yields (between 80 and 90%), and this is probably due to acyl migration of sn-2 to sn-1, which occurs spontaneously in glycerides [108].

## Prospective Aspects

### Whole Cells

For reduction of enzymatic processes costs, some researchers have studied microbial immobilization, such as fungal mycelia, bacteria, and yeasts cells, for their use as whole cell catalysts, taking advantage of functional proteins at the cell surface. From all whole cell support immobilization techniques, the most used is that named porous biomass support particles (BSPs) because it does not require chemical additives or cell preproduction; aseptic handling is unnecessary; higher enzyme production and rate of substrate mass transfer within BSP; the particles are reusable and resistant to mechanical shearing; the bioreactor scale-up is easy and presents lower costs compared to bioreactors used in other methods [8, 104].

The first example for biodiesel production using whole cell as biocatalysts was the Rhizopus oryzaemycelium immobilized in polyurethane foam [109]. The growth conditions were optimized regarding the production of intracellular lipase, as well as pretreatment methods and water content during methanolysis. The addition of substrates (olive oil or oleic acid) to the culture medium significantly improved lipase activity of the whole cell catalyst. The results for the obtainment of methyl esters of soybean oil using this catalyst at 32°C for 72 h (80–90% yield), when the addition of methanol to the system was implemented intermittently in the presence of 10–20% water, were very similar to those described with the use of extracellular lipases. Aiming to stabilize the R. oryzae cells, it was tested a cross-linking treatment with 0.1% glutaraldehyde, keeping the lipase dosage unaltered for 6 cycles. The yield of methyl ester varied between 70% and 83%, along 72 h of experiment. Without this treatment, the lipase activity decreased reaching a yield of 50% in the sixth batch [8, 91, 104].

In order to achieve higher yields of biodiesel using cells immobilized in BSP, Soetaert and Vandamme [104] used fixed bed systems. To increase the interfacial area between the reaction mixture and the whole cells, the former was emulsified by sonication before each batch cycle. When a gradual addition of methanol (ratio 4/1, methanol/oil) was conducted at a flow rate of $25\,Lh^{-1}$, a yield of 90% was obtained and maintained at about 80% for 10 batches. The authors attributed this decrease to the cell removal from the BSP, since it was detected a decrease of 56% in the cell concentration in the BSP between the first and the tenth batches.

As further examples, other freeze-dried whole cells, such as from R. chinensis mycelium, S. cerevisiae(containing intracellular R. oryzae lipase, ROL), and recombinant S. cerevisae expressing the lipase gene ofR. oryzae IFO 4697 (cell surface ROL), have been used as biocatalysts for the production of methyl biodiesel from soybean oil. In the absence of solvent, the yields observed for the cited examples were 86, 71, and 78%, respectively, after 165 h of reaction at 37°C [8, 102].

Salum et al. [110] showed that it is possible to decrease the costs associated to the synthesis of enzyme-catalyzed biodiesel, by using the fermented solids produced by cultivating Burkholderia cepacia LTEB11 on a mixture of sugarcane bagasse and sunflower seed meal. The authors used this fermented solids to catalyze the ethanolysis of soybean oil aiming to produce biodiesel in a fixed-bed reactor in a cosolvent-free system. They achieved 95% conversion after 46 h of reaction.

Although the use of whole cells does not require many of the steps related to the downstream process of biodiesel production, such as the isolation and purification of the enzyme after fermentation, processes in which the transesterification reaction is done by using immobilized enzymes or cells present at least one notable difference, which is the reaction time. Fukuda et al. [111] reported the use of Novozym 435 in a continuous process, for 7 h, and they observed yields of 92–94% in terms of methyl esters. When a fed-batch operation was considered, the yield was 87%, after 3.5 h of reaction with methyl oleate. The authors also compared the process performance by using whole cells in a fed-batch operation mode, and observed the necessity of 70 h of reaction in order to obtain yields of 80–90% of biodiesel. In the last case, the utilization of t-butanol as solvent would possibly reduce the reaction time, due to higher efficiency of mass transfer in the system.

## Use of Acid Oils

Some waste oils, by-products from vegetable oils processing, may also be suitable raw materials for biodiesel production. These oils usually present high contents of FFAs, and some examples are the sunflower oil and corn oil, which have 55.6% and 75.3% of FFAs, respectively. Pandey [91] reported the esterification and transesterification of these oils in the presence of hexane using immobilized lipase from C. Antarctica and observed yields of 64% and 50% of methyl esters, while maintaining the lipase stable for over 100 cycles.

Hou and Shaw [52] reported that the esterification of acid oils is much faster than the transesterification of nonacid oils. In the former case, it was necessary only 3 h of reaction and 1% of lipase for the esterification of FFAs, where a yield of 95% was achieved, whereas for the latter case, the same yield was observed only after 30 h of methanolysis and using a higher enzyme dosage (4%). One disadvantage of the esterification reaction is the formation of water as a by-product, which often inhibits the reaction of triglycerides. One possible solution to this is to conduct the reaction in two separate stages: first, esterification of the FFAs in the mixture, with the evaporation of the generated water; then the methanolysis of the triglycerides. In the first step, the molar ratio of methanol/FFA should be low, such as 1/2 and low quantity of enzyme (0.5% w/w) is needed. In the second step, on the other hand, the molar ratio between methanol and triglycerides should be changed to 1/1, and the enzyme quantity should be increased to about 6% (w/w).

## Hydro Esterification and Hybrid Catalysis

Hydro esterification is a process that combines two basic processes, the hydrolysis of triglycerides and the esterification of fatty acids, in sequential reactions, in order to produce biodiesel.

Talukder et al. [112] studied the use of residual cooking oil for biodiesel production by enzymatic hydrolysis accompanied by chemical esterification. The C. rugosa lipase used completely hydrolyzed the oil after 10 h of reaction. The FFAs were converted into biodiesel by chemical esterification using Amberlyst 15 (acidic styrene divinylbenzene), and a 99% conversion into biodiesel was obtained after 2 h. In this work, there was a loss of enzyme activity, and the hydrolysis yield was decreased to 92% after five batch cycles.

Cavalcanti-Oliveira et al. [113] studied the use of a lipase from Thermomyces lanuginosus (TL 100 L) in the hydrolysis of soybean oil in a hydroesterification process. The lipase hydrolyzed 89% of the oil after 48 h of reaction. This stage was followed by the esterification of the FFAs with methanol, which was catalyzed by niobic acid in pellets. They obtained 92% conversion of the FFAs into fatty acid methyl esters after only 1 h of incubation. Sousa et al. [114] studied the lipase from jatropha seeds for the hydrolysis of different raw materials for biodiesel production using hydro esterification strategy. The best conversions were obtained using soybean oil and jatropha oil, obtaining up to 98% of FFA after 2 h. The esterification of the FFAs from the jatropha oil with methanol was catalyzed by niobic acid in pellets, obtaining up to 97% conversion into biodiesel after 2 h. The biodiesel obtained from this process fulfilled all the legal requirements for its commercial use.

# CONCLUDING REMARKS

The use of enzyme catalysts (lipases) in biodiesel production is being increasingly studied because of the advantages that these catalysts present over chemically catalyzed and no catalytic processes. Some of the advantages offered by the use of lipases are lower energy consumption; lower thermal degradation of substrates and products; versatility in the use of raw materials, including possibility to use acid oils without the decrease of process efficiency; easier purification of the alkyl esters (biodiesel) and separation of the coproduct (glycerol), especially if immobilized enzymes or whole cells are used; environmental benefits, due to biodegradability of the catalyst.

Nevertheless, some process conditions should be taken into account in order to have a feasible enzyme-based technology for the production of biodiesel, and these include the establishment of descriptive correlations between the enzyme dosage and the substrate source, in order to rationalize enzyme usage and optimize costs [106]; deep study of reaction conditions and their optimization; the selection of a proper biocatalyst which can be reused and maintain its stability over several cycles; product recovery strategies, especially when a cosolvent is used in the reaction.

The enzyme-based production of biodiesel is still under development, and it seems that there is a tendency for the use of conventional technologies as a new application for lipases, such as their immobilization in magnetic nanoparticles [115], microwave and ultrasound-assisted transesterification [116], esterification in pressurized fluids [117], and transesterification in supercritical fluids [116]. Although technical aspects of such strategies may lead to conversion improvement, economic considerations must be investigated in more details.

# REFERENCES

1.  F. D. Gunstone and F. B. Padley, Lipid Technologies and Applications, Marcel Dekker, New York, NY, USA, 1997.

2.  J. J. Salas, J. Sánchez, U. S. Ramli, A. M. Manaf, M. Williams, and J. L. Harwood, "Biochemistry of lipid metabolism in olive and other oil fruits," Progress in Lipid Research, vol. 39, no. 2, pp. 151–180, and 2000.

3.  F. D. Gunstone, Vegetable Oils in Food Technology: Composition, Properties and Uses, Blackwell Publishing, Oxford, UK, 2002.

4.  M. F. Ali, B. M. E. Ali, and J. G. Speight, Handbook of Industrial Chemistry—Organic Chemicals, McGraw-Hill, New York, NY, USA, 2005.

5.  P. J. Eastmond and I. A. Graham, "Re-examining the role of the glyoxylate cycle in oilseeds," Trends in Plant Science, vol. 6, no. 2, pp. 72–77, 2001

6.  Food and Agriculture Organization of the United Nations (FAO), Food Outlook, FAO, New York, NY, USA, 2010.

7.  J. C. Melo, J. C. Teixeira, J. Z. Brito, J. G. A. Pacheco, and L. Stragevitch, "Produção de Biodiesel de Óleo de Oiticica," in Proceedings of the I Congresso da Rede Brasileira de Tecnologia do Biodiesel, pp. 164–169, August 2006.

8.  M. J. Dabdoub, J. L. Bronze, and M. A. Rampin, "Biodiesel: visão crítica do status atual e perspectivas na academia e na indústria," Química Nova, vol. 32, no. 3, pp. 776–792, 2009.

9.  F. D. Gunstone, J. L. Harwood, and A. J. Dijkstra, The Lipid Handbook, CRC Press, Boca Raton, Fla, USA, 2007.

10. L. M. P. Santos, "Nutritional and ecological aspects of buriti or aguaje (Mauritia flexuosa Linnaeus filius): a carotene-rich palm fruit from Latin America," Ecology of Food and Nutrition, vol. 44, no. 5, pp. 345–358, 2005

11. F. Danieli, O óleo de abacate (Persea americana Mill) como matéria-prima para indústria alimentícia, Dissertation. Thesis, University of São Paulo, Piracicaba, Brazil, 2006.

12. Y. M. Choo, "Palm oil carotenoids," Food and Nutrition Bulletin, vol. 15, no. 2, pp. 130–137, 1994.

13. M. C. T. Damaso, M. A. Passianoto, S. C. Freitas, D. M. G. Freire, R. C. A. Lago, and S. Couri, "Utilization of agroindustrial residues for lipase production by solid-state fermentation," Brazilian Journal of Microbiology, vol. 39, no. 4, pp. 676–681, 2008.

14. E. Wolski, E. Menusi, M. Mazutti, et al., "Response surface methodology for optimization of lipase production by an immobilized newly isolated Penicillium sp," Industrial & Engineering Chemistry Research, vol. 47, no. 23, pp. 9651–9657, 2008.

15. V. K. Garlapati, P. R. Vundavilli, and R. Banerjee, "Evaluation of lipase production by genetic algorithm and particle swarm optimization and their comparative study," Applied Biochemistry and Biotechnology, vol. 162, no. 5, pp. 1350–1361, 2010.

16. M. Singh, K. Saurav, N. Srivastava, and K Kannabiran, "Lipase production by Bacillus subtilis OCR-4 in solid state fermentation using ground nut oil cakes as substrates," Current Research Journal of Biological Sciences, vol. 2, no. 4, pp. 241–245, 2010.

17. T. F. C. Salum, P. Villeneuve, B. Barea, et al., "Synthesis of biodiesel in column fixed-bed bioreactor using the fermented solid produced by Burkholderia cepacia LTEB11," Process Biochemistry, vol. 45, no. 8, pp. 1348–1354, 2010

18. S. Y. Sun and Y. Xu, "Solid-state fermentation for "whole-cell synthetic lipase" production fromRhizopus chinensis and identification of the functional enzyme," Process Biochemistry, vol. 43, no. 2, pp. 219–224, 2008.

19. R. Bussamara, A. M. Fuentefria, E. S. D. Oliveira, et al., "Isolation of a lipase-secreting yeast for enzyme production in a pilot-plant scale batch fermentation," Bioresource Technology, vol. 101, no. 1, pp. 268–275, 2010.

20. S. Kumar, N. Katiyar, P. Ingle, and S. Negi, "Use of evolutionary operation (EVOP) factorial design techniques to develop a bioprocess using grease waste as substrate for lipase production," Bioresource Technology, vol. 102, no. 7, pp. 4909–4912, 2011.

21. Q. Li, W. Du, and D. Liu, "Perspectives of microbial oils for biodiesel production," Applied Microbiology and Biotechnology, vol. 80, no. 5, pp. 749–756, 2008.

22. M. G. Morais and J. A. V. Costa, "Fatty acids profile of microalgae cultived with carbon dioxide," Ciência e Agrotecnologia, vol. 32, no. 4, pp. 1245–1251, 2008.

23. G. Knothe, J. V. Gerpen, and J. Krahl, The Biodiesel Handbook, AOCS Press, Champaign, Ill, USA, 2005.

24. Y. B. Zuo, A. W. Zeng, X. G. Yuan, and K. T. Yu, "Extraction of soybean isoflavones from soybean meal with aqueous methanol modified supercritical carbon dioxide," Journal of Food Engineering, vol. 89, no. 4, pp. 384–389, 2008.

25. L. N. Ceci, D. T. Constenla, and G. H. Crapiste, "Oil recovery and lecithin production using water degumming sludge of crude soybean oils," Journal of the Science of Food and Agriculture, vol. 88, no. 14, pp. 2460–2466, 2008.

26. U. T. Bornscheuer, "Microbial carboxyl esterases: classification, properties and application in biocatalysis," FEMS Microbiology Reviews, vol. 26, no. 1, pp. 73–81, 2002.

27. K. E. Jaeger, B. W. Dijkstra, and M. T. Reetz, "Bacterial biocatalysts: molecular biology, three-dimensional structures, and biotechnological applications of lipases," Annual Review of Microbiology, vol. 53, pp. 315–351, 1999.

28. R. Sharma, Y. Chisti, and U. C. Banerjee, "Production, purification, characterization, and applications of lipases," Biotechnology Advances, vol. 19, no. 8, pp. 627–662, 2001.

29. H. F. Castro, A. A. Mendes, and J. C. Santos, "Modificação de óleos e gorduras por biotransformação," Química Nova, vol. 27, no. 1, pp. 146–156, 2004.

30. D. M. G. Freire and L. R. Castilho, "Lipases em Biocatálise," in Enzimas em Biotecnologia. Produção, Aplicações e Mercado, vol. 1, pp. 369–385, Interciência, Rio de Janeiro, Brazil, 2008.

31.  K. E. Jaeger and T. Eggert, "Lipases for biotechnology," Current Opinion in Biotechnology, vol. 13, no. 4, pp. 390–397, 2002.

32.  P. Villeneuve, "Plant lipases and their applications in oils and fats modification," European Journal of Lipid Science and Technology, vol. 105, no. 6, pp. 308–317, 2003.

33.  E. D. C. Cavalcanti, F. M. Maciel, P. Villeneuve, R. C. A. Lago, O. L. T. Machado, and D. M. G. Freire, "Acetone powder from dormant seeds of Ricinus communis L: lipase activity and presence of toxic and allergenic compounds," Applied Biochemistry and Biotechnology, vol. 137–140, no. 1–12, pp. 57–65, 2007.

34.  N. López, M. A. Pernas, L. M. Pastrana, A. Sánchez, F. Valero, and M. L. Rua, "Reactivity of pureCandida rugosa lipase isoenzymes (Lip1, Lip2, and Lip3) in aqueous and organic media. Influence of the isoenzymatic profile on the lipase performance in organic media," Biotechnology Progress, vol. 20, no. 1, pp. 65–73, 2004.

35.  J. Pfeffer, S. Richter, J. Nieveler, et al., "High yield expression of lipase A from Candida antarctica in the methylotrophic yeast Pichia pastoris and its purification and characterisation," Applied Microbiology and Biotechnology, vol. 72, no. 5, pp. 931–938, 2006.

36.  M. L. M. Fernandes, N. Krieger, A. M. Baron, P. P. Zamora, L. P. Ramos, and D. A. Mitchell, "Hydrolysis and synthesis reactions catalysed by Thermomyces lanuginosa lipase in the AOT/Isooctane reversed micellar system," Journal of Molecular Catalysis B, vol. 30, no. 1, pp. 43–49, 2004.

37.  Z.-Y. Shu, J.-K. Yang, and Y.-J. Yan, "Purification and characterization of a lipase from Aspergillus niger F044," Chinese Journal of Biotechnology, vol. 23, no. 1, pp. 96–101, 2007.

38.  H. Ogino, S. Nakagawa, K. Shinya, et al., "Purification and characterization of organic solvent-stable lipase from organic solvent-tolerant Pseudomonas aeruginosa LST-03," Journal of Bioscience and Bioengineering, vol. 89, no. 5, pp. 451–457, 2000.

39.  E. Lesuisse, K. Schanck, and C. Colson, "Purification and preliminary characterization of the extracellular lipase of Bacillus subtilis 168, an extremely basic pH-tolerant enzyme," European Journal of Biochemistry, vol. 216, no. 1, pp. 155–160, 1993.

40. J. F. M. Burkert, F. Maugeri, and M. I. Rodrigues, "Optimization of extracellular lipase production byGeotrichum sp. using factorial design," Bioresource Technology, vol. 91, no. 1, pp. 77–84, 2004.

41. R. R. Maldonado, Produção, purificação e caracterização da lipase de Geotrichum candidum obtida a partir de meios industriais, Disseration. thesis, University of Campinas, Campinas, Brazil, 2006.

42. M. Abrimic, I. Lescic, T. Korica, L. Vitale, W. Saenger, and J. Pigac, "Purification and properties of extracellular lipase from Streptomyces rimosus," Enzyme and Microbial Technology, vol. 25, no. 6, pp. 522–529, 1999.

43. J. Destain, D. Roblain, and P. Thonart, "Improvement of lipase production from Yarrowia lipolytica," Biotechnology Letters, vol. 19, no. 2, pp. 105–107, 1997.

44. A. Aloulou, J. A. Rodriguez, D. Puccinelli, et al., "Purification and biochemical characterization of the LIP2 lipase from Yarrowia lipolytica," Biochimica et Biophysica Acta, vol. 1771, no. 2, pp. 228–237, 2007.

45. A. I. S. Brigida, P. F. F. Amaral, L. R. B. Gonçalves, and M. A. Z. Coelho, "Characterization of an Extracellular lipase from Yarrowia lipolytica," in Proceedings of the European Congress of Chemical Engineering 6, (ECCE ‹07), vol. 2, Norhaven Book, Copenhagen, Denmark, September 2007.

46. H. Uhlig, Industrial Enzymes and Their Applications, John Wiley & Sons, New York, NY, USA, 1998.

47. H. Abbas, A. Hiol, V. Deyris, and L. Comeau, "Isolation and characterization of an extracellular lipase from Mucor sp. strain isolated from palm fruit," Enzyme and Microbial Technology, vol. 31, no. 7, pp. 968–975, 2002.

48. T. Godfrey and S. West, Industrial Enzymology, the Macmillan Press, London, UK, 1996.

49. P. J. Eastmond, "Cloning and characterization of the acid lipase from castor beans," Journal of Biological Chemistry, vol. 279, no. 44, pp. 45540–45545, 2004.

50. E. D. C. Cavalcanti, Avaliação da atividade lipásica da semente dormente de Ricinus communis, Disseration. Thesis, Federal University of Rio de Janeiro, Rio de Janeiro, Brazil, 2006.

51.  R. R. Costa Neto, Obtenção de ésteres alquilicos (Biodiesel) por via enzimática a partir de óleo de soja, Ph.D. thesis, Federal University of Santa Catarina, Florianópolis, Brazil, 2002.

52.  C. T. Hou and J. F. Shaw, Biocatalysis and Bioenergy, John Wiley & Sons, Hoboken, NJ, USA, 2008.

53.  A. Nag, Biofuels Refining and Performance, McGraw-Hill, New York, NY, USA, 2008.

54.  L. R. Castilho, C. M. S. Polato, E. A. Baruque, G. L. Sant›Anna Jr., and D. M. G. Freire, "Economic analysis of lipase production by Penicillium restrictum in solid-state and submerged fermentations,"Biochemical Engineering Journal, vol. 4, no. 3, pp. 239–247, 2000.

55.  L. A. I. Azeredo, P. M. Gomes, G. L. Sant›Anna Jr., L. R. Castilho, and D. M. G. Freire, "Production and regulation of lipase activity from Penicillium restrictum in submerged and solid-state fermentations," Current Microbiology, vol. 54, no. 5, pp. 361–365, 2007.

56.  M. B. Palma, A. L. Pinto, A. K. Gombert, et al., "Lipase production by Penicillium restrictum using solid waste of industrial babassu oil production as substrate," Applied Biochemistry and Biotechnology, vol. 84–86, pp. 1137–1145, 2000.

57.  M. L. E. Gutarra, E. D. C. Cavalcanti, D. M. G. Freire, L. R. Castilho, and G. L. Sant›Anna Jr., "Lipase production by solid-state fermentation: cultivation conditions and operation of tray and packed-bed bioreactors," Applied Biochemistry and Biotechnology, vol. 121, no. 1–3, pp. 105–116, 2005.

58.  M. L. E. Gutarra, M. G. Godoy, F. Maugeri, M. I. Rodrigues, D. M. G. Freire, and L. R. Castilho, "Production of an acidic and thermostable lipase of the mesophilic fungus Penicillium simplicissimumby solid-state fermentation," Bioresource Technology, vol. 100, no. 21, pp. 5249–5254, 2009.

59.  E. D. C. Cavalcanti, M. L. E. Gutarra, D. M. G. Freire, L. R. Castilho, and G. L. Sant›Anna Jr., "Lipase production by solid-state fermentation in fixed-bed bioreactors," Brazilian Archives of Biology and Technology, vol. 48, pp. 79–84, 2005.

60.  M. Diluccio, F. Capra, N. P. Ribeiro, G. D. L. P. Vargas, D. M. G. Freire, and D. Oliveira, "Evaluation of lipase production by

solid state fermentation by Penicillium simplicissimum using soy cake,"Applied Biochemistry and Biotechnology, vol. 113, pp. 173–180, 2004.

61.  M. G. Godoy, M. L. E. Gutarra, A. M. Castro, O. L. M. Tavares, and D. M. G. Freire, "Adding value to a toxic residue from the biodiesel industry: production of two distinct pool of lipases from Penicillium simplicissimum in castor bean waste," Journal of Industrial Microbiology & Biotechnology. In press.

62.  P. V. Rao, K. Jayaraman, and C. M. Lakshmanan, "Production of lipase by Candida rugosa in solid state fermentation. 1: determination of significant process variables," Process Biochemistry, vol. 28, no. 6, pp. 385–389, 1993.

63.  J. A. Rodriguez, J. C. Mateos, J. Nungaray, et al., "Improving lipase production by nutrient source modification using Rhizopus homothallicus cultured in solid state fermentation," Process Biochemistry, vol. 41, no. 11, pp. 2264–2269, 2006.

64.  J. C. M. Mateos Diaz, J. A. Rodríguez, S. Roussos, et al., "Lipase from the thermotolerant fungusRhizopus homothallicus is more thermostable when produced using solid state fermentation than liquid fermentation procedures," Enzyme and Microbial Technology, vol. 39, no. 5, pp. 1042–1050, 2006.

65.  N. D. Mahadik, U. S. Puntambekar, K. B. Bastawde, J. M. Khire, and D. V. Gokhale, "Production of acidic lipase by Aspergillus niger in solid state fermentation," Process Biochemistry, vol. 38, no. 5, pp. 715–721, 2002.

66.  N. R. Kamini, J. G. S. Mala, and R. Puvanakrishnan, "Lipase production from Aspergillus niger by solid-state fermentation using gingelly oil cake," Process Biochemistry, vol. 33, no. 5, pp. 505–511, 1998.

67.  J. Cordova, M. Nemmaoui, M. Ismaïli-Alaoui, et al., "Lipase production by solid state fermentation of olive cake and sugar cane bagasse," Journal of Molecular Catalysis. B, vol. 5, no. 1–4, pp. 75–78, 1998.

68.  I. ul-Haq, S. Idrees, and M. I. Rajoka, "Production of lipases by Rhizopus oligosporous by solid-state fermentation," Process Biochemistry, vol. 37, no. 6, pp. 637–641, 2002.

69. D. A. Mitchell, B. K. Losane, A. Durand, et al., "General principles of reactors design and operation for SSC," in Solid Substrate Cultivation, H. Doelle, D. A. Mitchell, and C. E. Rols, Eds., pp. 115–139, Elsevier Applied Science, Amsterdam, The Netherlands, 1992.

70. D. A. Mitchell, A. Pandey, P. Sangsurasak, and N. Krieger, "Scale-up strategies for packed-bed bioreactors for solid-state fermentation," Process Biochemistry, vol. 35, no. 1-2, pp. 167–178, 1999.

71. D. A. Mitchell, N. Krieger, D. M. Stuart, and A. Pandey, "New developments in solid-state fermentation II. Rational approaches to the design, operation and scale-up of bioreactors," Process Biochemistry, vol. 35, no. 10, pp. 1211–1225, 2000.

72. D. A. Mitchell, M. Berovic, and N. Krieger, "Overview of solid state bioprocessing," Biotechnology Annual Review, vol. 8, pp. 183–225, 2002.

73. F. D. H. Dalsenter, G. Viccini, M. C. Barga, D. A. Mitchell, and N. Krieger, "A mathematical model describing the effect of temperature variations on the kinetics of microbial growth in solid-state culture," Process Biochemistry, vol. 40, no. 2, pp. 801–807, 2005.

74. P. Sangsurasak and D. A. Mitchell, "Incorporation of death kinetics into a 2-dimensional dynamic heat transfer model for solid state fermentation," Journal of Chemical Technology and Biotechnology, vol. 64, no. 3, pp. 253–260, 1995.

75. M. T. Hardin, D. A. Mitchell, and T. Howes, "Approach to designing rotating drum bioreactors for solid-state fermentation on the basis of dimensionless design factors," Biotechnology and Bioengineering, vol. 67, no. 3, pp. 274–282, 2000.

76. M. A. I. Schutyser, P. Pagter, F. J. Weber, W. J. Briels, R. M. Boom, and A. Rinzema, "Substrate aggregation due to aerial hyphae during discontinuously mixed solid-state fermentation withAspergillus oryzae: experiments and modeling," Biotechnology and Bioengineering, vol. 83, no. 5, pp. 503–513, 2003.

77. M. M. Santos, A. S. Rosa, S. Dal›Boit, D. A. Mitchell, and N. Krieger, "Thermal denaturation: is solid-state fermentation really a good technology for the production of enzymes?" Bioresource Technology, vol. 93, no. 3, pp. 261–268, 2004.

78.  A. Aloulou, J. A. Rodriguez, D. Puccinelli, et al., "Purification and biochemical characterization of the LIP2 lipase from Yarrowia lipolytica," Biochimica et Biophysica Acta, vol. 1771, no. 2, pp. 228–237, 2007.

79.  M. L. Rua, C. Schmidt-Dannert, S. Wahl, A. Sprauer, and R. D. Schmid, "Thermoalkalophilic lipase of Bacillus thermocatenulatus: large-scale production, purification and properties: aggregation behaviour and its effect on activity," Journal of Biotechnology, vol. 56, no. 2, pp. 89–102, 1997.

80.  M. Teissere, M. Borel, B. Caillol, J. Nari, A. M. Gardies, and G. Noat, "Purification and characterization of a fatty acyl-ester hydrolase from post-germinated sunflower seeds," Biochimica et Biophysica Acta, vol. 1255, no. 2, pp. 105–112, 1995.

81.  H. Sztajer, H. Lünsdorf, H. Erdmann, U. Menge, and R. Schmid, "Purification and properties of lipase from Penicillium simplicissimum," Biochimica et Biophysica Acta, vol. 1124, no. 3, pp. 253–261, 1992.

82.  K. Isobe, K. Nokihara, S. Yamaguchi, T. Mase, and R. D. Schmid, "Crystallization and characterization of monoacylglycerol and diacylglycerol lipase from Penicillium camembertii," European Journal of Biochemistry, vol. 203, no. 1-2, pp. 233–237, 1992.

83.  A. G. Cunha, G. F. Lorente, M. L. E. Gutarra, et al., "Separation and immobilization of lipase from Penicillium simplicissimum by selective adsorption on hydrophobic supports," Applied Biochemistry and Biotechnology, vol. 156, no. 1–3, pp. 133–145, 2009.

84.  R. K. Saxena, A. Sheoran, B. Giri, and W. S. Davidson, "Purification strategies for microbial lipases," Journal of Microbiological Methods, vol. 52, no. 1, pp. 1–18, 2003.

85.  P. D. María, J. M. Sánchez-Montero, J. V. Sinisterra, and A. R. Alcântara, "Understanding Candida rugosa lipases: an overview," Biotechnology Advances, vol. 24, no. 2, pp. 180–196, 2006.

86.  P. F. F. Amaral, Produção de lipase de Yarrowia lipolytica em biorreator multifásico, Ph.D. thesis, Federal University of Rio de Janeiro, Rio de Janeiro, Brazil, 2007.

87. F. V. P. Meirelles, Produção de lipase de Yarrowia lipolytica (IMUFRJ50682), Ph.D. thesis, Federal University of Rio de Janeiro, Rio de Janeiro, Brazil, 1997.

88. M. F. C. P. J. S. Gama, Produção e caracterização de lipases de Penicillium restrictum, Dissertation. thesis, Federal University of Rio de Janeiro, Rio de Janeiro, Brazil, 2000.

89. T. S. M. Martins, Produção e purificação de lipases de Yarrowia lipolytica (IMUFRJ50682), Dissertation. thesis, Federal University of Rio de Janeiro, Rio de Janeiro, Brazil, 2001.

90. R. V. Almeida, Clonagem, expressão, caracterização e modelagem estrutural de uma esterase termoestável de Pyrococcus furiosus, Ph.D. thesis, University of Rio de Janeiro, Rio de Janeiro, Brazil, 2005.

91. A. Pandey, Handbook of Plant-Based Biofuels, CRC Press, Boca Raton, Fla, USA, 2009.

92. P. A. Z. Suarez, A. L. F. Santos, J. P. Rodrigues, and M. B. Alves, "Biocombustíveis a partir de óleos e gorduras: desafios tecnológicos para viabilizá-los," Química Nova, vol. 32, no. 3, pp. 768–775, 2009.

93. C. M. Drapcho, N. P. Nhuan, and T. H. Walker, Biofuels Engineering Process Technology, McGraw-Hill, New York, NY, USA, 2008.

94. R. C. Rodrigues, Síntese de biodiesel através de transesterificação enzimática de óleos vegetais catalisada por lipase imobilizada por ligação covalente multipontual, Ph.D. thesis, Federal University of Rio Grande do Sul, Porto Alegre, Brazil, 2009.

95. M. Sadeghi, "Purification of soybean oil with phospholipase A1," Theoretical and Experimental Chemistry, vol. 46, no. 2, pp. 132–137, 2010.

96. B. Yang, R. Zhou, J. G. Yang, Y. H. Wang, and W. F. Wang, "Insight into the enzymatic degumming process of soybean oil," Journal of the American Oil Chemists Society, vol. 85, no. 5, pp. 421–425, 2008.

97. K. Clausen, "Enzymatic oil-degumming by a novel microbial phospholipase," European Journal of Lipid Science and Technology, vol. 103, no. 6, pp. 333–340, 2001.

98. L. De Maria, J. Vind, K. M. Oxenboll, A. Svendsen, and S. Patkar, "Phospholipases and their industrial applications," Applied Microbiology and Biotechnology, vol. 74, no. 2, pp. 290–300, 2007.

99. S. K. Roy, B. V. S. K. Rao, and R. B. N. Prasad, "Enzymatic degumming of rice bran oil," Journal of the American Oil Chemist›s Society, vol. 79, no. 8, pp. 845–846, 2002.

100. T. Tan, J. Lu, K. Nie, L. Deng, and F. Wang, "Biodiesel production with immobilized lipase: a review,"Biotechnology Advances, vol. 28, no. 5, pp. 628–634, 2010.

101. P. M. Nielsen, J. Brask, and L. Fjerbaek, "Enzymatic biodiesel production: technical and economic considerations," European Journal of Lipid Science and Technology, vol. 110, no. 8, pp. 692–700, 2008.

102. A. Robles-Medina, P. A. Gonzalez-Moreno, L. Esteban-Cerdan, and E. Molina-Grima, "Biocatalysis: towards ever greener biodiesel production," Biotechnology Advances, vol. 27, no. 4, pp. 398–408, 2009.

103. W. Du, W. Li, T. Sun, X. Chen, and D. Liu, "Perspectives for biotechnological production of biodiesel and impacts," Applied Microbiology and Biotechnology, vol. 79, no. 3, pp. 331–337, 2008.

104. W. Soetaert and E. J. Vandamme, Biofuels, John Wiley & Sons, Hoboken, NJ, USA, 2009.

105. S. V. Ranganathan, S. L. Narasimhan, and K. Muthukumar, "An overview of enzymatic production of biodiesel," Bioresource Technology, vol. 99, no. 10, pp. 3975–3981, 2008.

106. C. Z. Brusamarelo, E. Rosset, A. Césaro, et al., "Kinetics of lipase-catalyzed synthesis of soybean fatty acid ethyl esters in pressurized propane," Journal of Biotechnology, vol. 147, no. 2, pp. 108–115, 2010.

107. A. L. F. Matassoli, I. N. S. Corrêa, M. F. Portilho, C. O. Veloso, and M. A. P. Langone, "Enzymatic synthesis of biodiesel via alcoholysis of palm oil," Applied Biochemistry and Biotechnology, vol. 155, no. 1–3, pp. 347–355, 2009.

108. Y.-D. Wang, X.-Y. Shen, Z.-L. Li, et al., "Immobilized recombinant Rhizopus oryzae lipase for the production of biodiesel in solvent

free system," Journal of Molecular Catalysis B, vol. 67, no. 1-2, pp. 45–51, 2010.

109. K. Ban, M. Kaieda, T. Matsumoto, A. Kondo, and H. Fukuda, "Whole cell biocatalyst for biodiesel fuel production utilizing Rhizopus oryzae cells immobilized within biomass support particles,"Biochemical Engineering Journal, vol. 8, no. 1, pp. 39–43, 2001.

110. T. F. C. Salum, P. Villeneuve, B. Barea, et al., "Synthesis of biodiesel in column fixed-bed bioreactor using the fermented solid produced by Burkholderia cepacia LTEB11," Process Biochemistry, vol. 45, no. 8, pp. 1348–1354, 2010.

111. H. Fukuda, S. Hama, S. Tamalampudi, and H. Noda, "Whole-cell biocatalysts for biodiesel fuel production," Trends in Biotechnology, vol. 26, no. 12, pp. 668–673, 2008.

112. M. M. R. Talukder, J. C. Wu, and L. P. L. Chua, "Conversion of waste cooking oil to biodiesel via enzymatic hydrolysis followed by chemical esterification," Energy and Fuels, vol. 24, no. 3, pp. 2016–2019, 2010.

113. E. D. Cavalcanti-Oliveira, P. R. R. Silva, A. P. Ramos, D. A. G. Aranda, and D. M. G. Freire, "Study of soybean oil hydrolysis catalyzed by Thermomyces lanuginosus lipase and its application to biodiesel production via hydroesterification," Enzyme Research, vol. 2011, Article ID 618692, 8 pages, 2011.

114. J. S. Sousa, E. D. Cavalcanti-Oliveira, D. A. G. Aranda, and D. M. G. Freire, "Application of lipase from the physic nut (Jatropha curcas L.) to a new hybrid (enzyme/chemical) hydroesterification process for biodiesel production," Journal of Molecular Catalysis B, vol. 65, no. 1–4, pp. 133–137, 2010.

115. W. Xie and N. Ma, "Enzymatic transesterification of soybean oil by using immobilized lipase on magnetic nano-particles," Biomass and Bioenergy, vol. 34, no. 6, pp. 890–896, 2010.

116. A. P. Vyas, J. L. Verma, and N. Subrahmanyam, "A review on FAME production processes," Fuel, vol. 89, no. 1, pp. 1–9, 2010.

117. G. Kuhn, M. Marangoni, D. M. G. Freire, et al., "Esterification activities of non-commercial lipases after pre-treatment in pressurized propane," Journal of Chemical Technology and Biotechnology, vol. 85, no. 6, pp. 839–844, 2010.

Chapter 4

# Sustainable Algae Biodiesel Production in Cold Climates

Rudras Baliga and Susan E. Powers

Center for the Environment, Clarkson University, 8 Clarkson Aveue, Potsdam, NY 13699-5700, USA

## ABSTRACT

This life cycle assessment aims to determine the most suitable operating conditions for algae biodiesel production in cold climates to minimize energy consumption and environmental impacts. Two hypothetical photobioreactor algae production and biodiesel plants located in Upstate New York (USA) are modeled. The photobioreactor is assumed to be housed within a greenhouse that is located adjacent to a fossil fuel or biomass power plant that can supply waste heat and flue gas containing $CO_2$ as a primary source of carbon. Model results show that the biodiesel areal productivity is high (19 to 25 L of $BD/m^2/yr$). The total life cycle energy consumption was between 15 and 23 MJ/L of algae BD and 20 MJ/L of soy BD. Energy consumption and air emissions for algae biodiesel are substantially lower than soy biodiesel when waste heat was utilized. Algae's most substantial contribution is a significant decrease in the petroleum consumed to make the fuel.

INTRODUCTION

In 1998, an amendment to the U.S. Energy Policy Act (EP Act) of 1992 triggered the rapid expansion of the US biodiesel industry. This act required that a fraction of new vehicles purchased by federal and state governments be alternative fuel vehicles. The U.S. Energy Independence and Security Act (EISA) of 2007 further mandated the production of renewable fuels to 36 billion gallons (136 billion liters) per year by 2022, including biodiesel. Crops such as soybeans and canola account for more than three quarters of all biodiesel feedstocks in the U.S. [1].

About 14% of U.S. soybean production and 4% of global soybean production were used by the U.S. biodiesel industry to produce fuel in 2007 [1]. The use of oil crops for fuel has been criticized because the expansion of biodiesel production in the United States and Europe has coincided with a sharp increase in prices for food grains and vegetable oils [2]. The production of biodiesel from feedstocks that do not use arable land can be accomplished either by using biomass that is currently treated as waste or by introducing a new technology that allows for the development of new feedstocks for biodiesel that utilize land that is unsuitable for food production.

Microalgae have the potential to displace other feedstocks for biodiesel owing to its high vegetable oil content and biomass production rates [3]. The vegetable oil content of algae can vary with growing conditions and species, but has been known to exceed 70% of the dry weight of algae biomass [4]. Microalgae could have significant social and environmental benefits because they do not compete for arable land with food crops and microalgae cultivation consumes less water than other crops [5]. Algae also grow in saline waters that are unsuitable for agricultural practices or consumption. This makes algae well suited for areas where cultivation of other crops is difficult [6, 7]. High biomass productivities may be achieved with indoor or outdoor photobioreactors (PBRs) [8]. In cold climates, PBRs have been used successfully, when housed within greenhouses and provided with artificial lighting

Microalgae biodiesel has received much attention in news media. Considerable progress has been made in the field of algae biomorphology [9–11]. In recent decades, however, little quantitative research has been done on the energy and environmental impacts of microalgae biodiesel production on a life cycle basis. The life cycle concept is a cradle to grave systems approach for the study of feedstocks, production, and use. The objective of this work was to assess the feasibility of algae biodiesel production in New York State (USA) based on life cycle energy and environmental impact parameters. Upstate NY was chosen as a challenging case for algae biodiesel production due to shorter days and cold temperatures during winter months. The productivity, energy consumption, and environmental emissions associated with the algae/BD production lifecycle were quantified in order to identify the best growing conditions and assess its impacts relative to soybean biodiesel.

# METHODOLOGY

## System Boundary and Scope

The life cycle concept is a cradle to grave systems approach for the study of feedstocks, production, and use. The concept revolves around the recognition of different stages of production starting from upstream use of energy to cultivation of the feedstock, followed by the different processing stages. A life cycle inventory assessment allows for the quantification of mass and energy streams such as energy consumption, material usage, waste production, and generation of coproducts. A summary of the sustainability assessment metrics used for this life cycle inventory of microalgae feedstock for biodiesel production is presented in Table 1.

**Table 1:** Life cycle sustainability metrics for biodiesel

| Environmental Impact | Sustainability Metrics |
|---|---|
| Energy and Resource Consumption | Total energy consumed (MJ/L BD) |
| | Fossil fuel energy consumed (MJ/L BD) |
| | Petroleum consumed (MJ/L BD) |
| | Land required (m²/L of BD) |
| | Water required (L water/L BD) |
| Climate Change | Net greenhouse gas emissions (g $CO_2$ equivalents/L of BD) |
| Acidification | Acidification potential (g $SO_2$eq./L BD) |
| Toxic Emissions | Particulate matter emissions ($PM_{10}$, $PM_{25}$) |
| | Carbon monoxide emissions |
| | Volatile organic carbon emissions |

Figure 1 provides an overview of the system boundary used in this analysis, which includes the production of algae and biodiesel via a transesterification reaction. The boundary includes all upstream mass and energy flows that are required to make the chemical and energy resources required for the processing. The production of biodiesel from algae and direct energy consumption is characterized by four distinct stages: cultivation, dewatering/drying, oil extraction, and transesterification (Figure 1). The energy consumed and subsequent emissions for fuel production, electricity generation, and chemical production comprise the upstream energy consumption and emissions. Biodiesel and algae meal are the products leaving the system boundary. The use of these products is not directly included within the analysis.

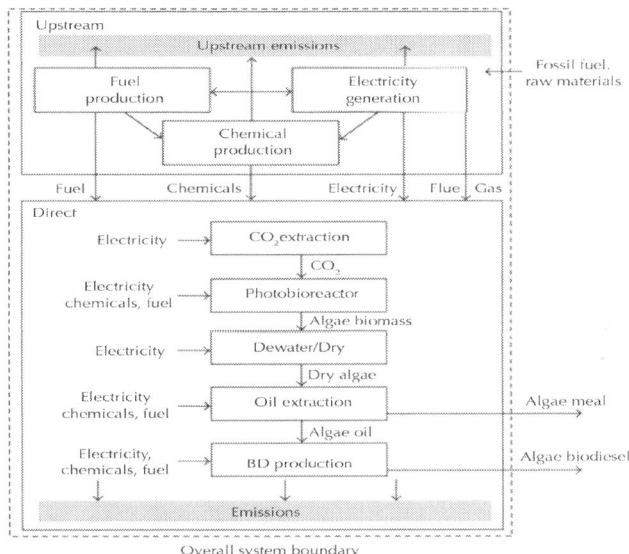

**Figure 1:** Flowchart depicting system boundary for life cycle inventory of biodiesel from microalgae.

The hypothetical algae and biodiesel production processing facilities considered are located in upstate placeStateNew York. The facilities are assumed to be adjacent to a biomass or fossil fuel electricity generation plant for access to the carbon dioxide in their flue gas and waste heat in order to maximize the utilization of waste resources within this system. Waste heat is considered to have no value as an energy product; so it is not counted as part of the total energy resources consumed by the facility.

Two different locations were considered for the microalgae biodiesel facility: Syracuse, NY ($43^0$ 2' N, $76^08'$ W) and Albany, NY ($42^07'$ N, $73^08'$ W). Although these locations are at approximately the same latitude and have very similar hours of daylight, the Syracuse area is colder and cloudier throughout the year due to its proximity to the Great Lakes. Albany offers more intense natural lighting and less severe winter temperatures (Figure 2). Three specific cases were considered for each of these locations:

1.  greenhouse structure to maximize natural lighting; natural gas used to maintain the system temperature;

2. greenhouse structure to maximize natural lighting; waste heat used to maintain the system temperature;

3. a well-insulated facility that allows for no natural lighting but requires substantially less heat.

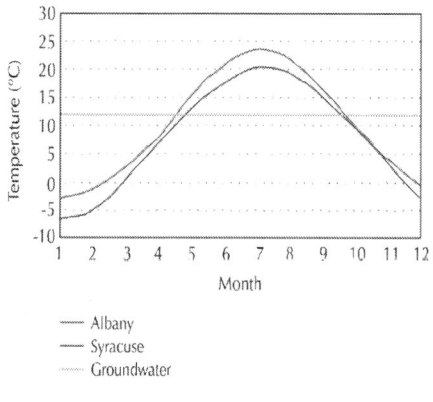

(a)

(b)

**Figure 2:** Monthly average temperature (a) and total monthly solar irradiance (b) for Syracuse, NY, and Albany, NY.

The PBRs are assumed to operate continuously, using artificial lighting when natural lighting is not sufficient. In all cases, it was assumed that Phaeodactylum tricornutum algae would be grown for biodiesel production. This algae species has a relatively high oil content (about 30% by dry weight), is resistant to contamination, and has been previously utilized to produce biodiesel [12, 13].

Estimating the environmental and energy lifecycle impacts requires quantification of the mass and energy flows through this system. A mathematical model for the algae production process was developed in the work presented here. As shown in Figure 3, the mass and energy flows estimated with the algae production model were used in conjunction with the Greenhouse gases, Related Emissions, and Energy use in Transportation (GREET) model 1.8a developed at the Argonne National Laboratories [14]. GREET provided the general framework and structure for the lifecycle inventory, especially aspects of the transesterification process and energy and emissions related to the upstream production of chemicals and energy resources. BD production from soybeans, which is used here as a benchmark for comparison, was taken directly from the GREET model. GREET is a widely accepted model and many studies and analyses have been based upon it because of its vast data on energy sources and the associated emissions (e.g., [15–17]). The default values for soybean production, oil extraction, and transesterification were taken as GREET default values [14], which are representative of the Midwestern region of the United States where most soybeans are grown. These were based initially on an LCA completed at the National Renewable Energy Lab [18] and updated to keep the GREET model as current as possible (e.g., [17]). There is only a small production of soybeans in NY State, with yields well below the average yield in the Midwest. Thus, no attempt was made to match the geographic system boundaries for biodiesel from algae to that of soybeans.

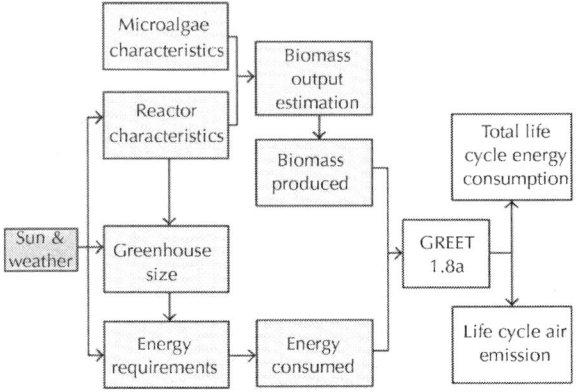

**Figure 3:** Overview of microalgae biomass LCA model. The yellow boxes represent the contributions of the work presented here. The soybean LCA results were taken almost entirely from GREET.

Uncertainty in the data was addressed by utilizing Monte Carlo simulations to input a range of values for parameters. For a given assumption or variable with a distribution as input, the commercially available software, Crystal Ball was utilized to determine a forecast or range of possible outputs. Standard error bars were created utilizing the mean value of the forecast and 95% certainty.

# Algae Production Models

The biomass production model utilizes solar data and a biological growth rate to estimate actual yields for algae biomass for a photobioreactor system [13, 19].

## *Microalgae Plant Setup*

Hypothetical tubular closed photobioreactors (PBRs) were modeled in this case to predict algae production and account for energy consumption and emissions in Syracuse and Albany NY. The PBR plant setup is illustrated in Figure 4. It was assumed that processes such as dewatering and transesterification could be carried out on site, thus eliminating the need for transportation.

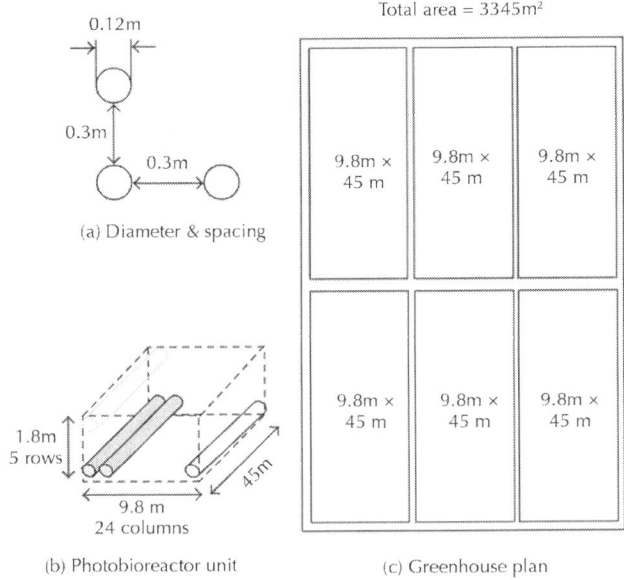

0.12m

0.3m

0.3m

(a) Diameter & spacing

1.8m
5 rows

9.8 m
24 columns

45m

(b) Photobioreactor unit

Total area = 3345m²

| 9.8m × 45 m | 9.8m × 45 m | 9.8m × 45 m |
| 9.8m × 45 m | 9.8m × 45 m | 9.8m × 45 m |

(c) Greenhouse plan

**Figure 4:** Photobioreactor system layout (not to scale).

The various dimensions and parameters for the PBR were taken from recommendations of previous studies in order to depict a realistic setup [12, 13]. The PBR setup was designed for a maximum of 30 hours detention time. The maximum effluent concentration ($C_e$) was fixed at 3.4 kg/m³ with a recycle ratio of 0.35 [13, 20] and an allowable superficial fluid velocity of 0.3 m/s [21]. Since a long tubular length is required to meet these constraints (32,400 m), the PBR is split up into 6 units each of which is 61 m³ (5,400 m long, 0.12 m diameter). Stacking of tubes reduces the total foot print area of the greenhouse. All tubes are connected and algae broth passes through all six units.

The floor area or foot print area of the greenhouse house was determined from the volume of the reactor and type of cultivation (annual/seasonal operation) and the specific processes. The diameter of tubes was set at 0.12 m for all cases since it is a widely reported size for PBRs [3, 13, 22, and 23]. The spacing of tubes was set at 0.3 m. This is an important factor since it defines the total floor size, which in turn influences heating and lighting requirements. The various parameters related to plant setup are summarized in Table 2.

**Table 2:** Summary of photobioreactor and greenhouse parameters

| Parameter | Value | Depends upon | Reference |
|---|---|---|---|
| Diameter of tubes | 0.12 m | Larger diameter pipes can cause cell shading | [3] |
| Spacing of tubes | 0.3 | Greater spacing is desirable to avoid shading | [13] |
| Flow rate (Q) | $3.4*10^{-3}$ m³/s | Algae species and tube diameter | [21] |
| Recycle Ratio (r) | 0.35 | | [20] |
| Total volume of reactor (V) | 366 m³ | Maximum residence time (30 hrs) | [12] |
| Influent concentration ($C_i$) | 1.2 kg/m³ | | |
| Max. Effluent concentration ($C_e$) | 3.4 kg/m³ | Growth rate, PBR set up | [13] |

## Estimating Biomass Output

Microalgae productivity is estimated from the location, reactor specifications, and microalgae data. It is assumed that $CO_2$ and nutrients are provided in excess to the microalgae culture through the media, thereby making light the only limiting factor for cell growth and decay [12]. If adequate lighting is available, the specific growth rate $\mu$ is determined from the average irradiance available $I_{avg} (\mu E / m^2 - s)$ [19]:

$$\mu = \mu_{max} \frac{I_{avg}^n}{K_I^n + I_{avg}^n},$$   (1)

where $K_I$ is the half saturation constant (i.e., $I_{avg}$ for which half of $\mu_{max}$ is attained), and the exponent n is a unitless empirical constant. Both $K_I$ and n are constant for a given species of algae. Note that decay of algae cells during the hours with light is incorporated into the maximum specific growth rate ($\mu_{max}$) (h$^{-1}$) since values provided by Molina Grima et al. [22] and Fernandez et al. [23] were determined from the net growth rate. $I_{avg}$ is determined from the Beer-Lambert equation:

$$I_{avg} = \frac{I}{\varphi_{eq}K_aC_i}[1-\exp(-\varphi_{eq}K_aC_i)], \tag{2}$$

where $C_i$ (kg/m$^3$) is the influent biomass concentration. The path length of light within the reactor is given by $\varphi_{eq}$, which is the ratio of the tube diameter to the cosine of the solar zenith angle. The photosynthetically active irradiance (I) ($\mu Em^{-2}s^{-1}$) is a function of various solar angles and the total solar irradiance. Hourly solar data were available from NREL's solar database [24], and thus, algae cell growth was determined at an interval of an hour. The analysis of the solar data to estimate I and $\varphi_{eq}$ is included in the appendix.

The PBR is modeled as a series of plug flow reactors where the effluent concentration of each reactor is the influent concentration for the next. It is assumed that steady-state conditions prevail for each hour since irradiance is taken as a constant over that time period. Utilizing a Monod reaction rate for substrate utilization [22], the resulting steady-state plug flow reactor equation for each segment can be written as

$$u\frac{dc}{dz} = \mu_{max}\frac{I_{avg}^n}{K_I^n + I_{avg}^n}C, \tag{3}$$

where C is the biomass concentration and u is the fluid velocity. Integrating this expression can provide the effluent concentration for each reactor segment "j" that represents one hour of residence time at an average irradiation rate for that hour:

$$\ln(\frac{C_{j+1}^k}{C_j^k}) = \mu_{max}\frac{(I_{avg(j+1)}^k)^n}{K_I^n + (I_{avg(j+1)}^k)^n}(\frac{L}{u}), \tag{4}$$

where j and j+1 indicate the start and end location along the reactor length of the one hour segment. $C_{j+1}$ is calculated in series to determine the reactor effluent concentration, $C_e$, for each hour k. The growth rate during that hour is defined by the average irradiance for that hour of the day. The total biomass produced per day $(M_{BM})(kgd^{-1})$ is estimated from the flow rate $(Q)(m^3 hr^{-1})$, recycle ratio (r), and the effluent concentration $C_e$ (kg m$^{-3}$):

$$M_{BM} = \sum_{k=1}^{24} Q(C_e^k)(1-r) \qquad (5)$$

The total microalgae biomass produced can be determined from (2), (4), and (5) along with algae growth parameters and solar irradiation data.

The temperature requirements for algae differ by species. In general, faster growing algae species favor higher media temperatures of about 20–3C [10]. The algae-related constants used for P. tricornutum in the model are included in Table 3. This species was selected because it has been used in the past to produce microalgae biodiesel and all relevant data were available [22].

**Table 3:** P. tricornutum growth parameters

| Parameter | Value | Reference |
|---|---|---|
| $\mu_{max}$ | 0.063 (h$^{-1}$) | [22] |
| B | 0.0018 (h$^{-1}$) | [25] |
| $K_l$ | 3426 ($\mu Em^{-2}s^{-1}$) | [22] |
| n | 1–1.34 | [25] |
| $E_f$ | $1.74 \pm 0.09 \mu EJ^{-1}$ | [22] |
| $X_p$ | 2–4% | [25] |
| Oil Content (%dry weight) | 30% | [13] |
| Water Content of algae | 40% | [13] |

| | | |
|---|---|---|
| Cultivation temperature | 25°C | [10]] |

# Energy Consumed during Microalgae Cultivation

Cultivating microalgae in closed systems is an energy intensive process, especially in regions with low temperatures and limited natural lighting [13]. The algae growth and harvesting stage involve a large number of intermediate processes for which estimates of the energy consumption were developed here. Energy consumption requirements for extraction and transesterification are already provided in the GREET model [14]. It was assumed that the processes for transesterification of algae oil are identical to that of soy oil. Thus, the chemical (methanol and sodium hydroxide) and energy consumption and the energy and emissions associated with their production were taken directly as the default parameters in GREET.

## *Water Heating*

Energy requirements for a natural gas water heating system were determined by using the specific heat of water and the efficiency rating of the heater as provided by the manufacturer (EF = 0.82 [26]). It was assumed that groundwater, at an initial temperature ($T_{inlet}$) of 12°C [27], would be heated to a thermostat set point ($T_t$) of 25°C.

## *Media Circulation*

Electric pumps are used to circulate the media through the entire length of the reactor. The input electrical power required to operate the pumps ($P_p$) (W) can be given by [28]

$$P_p = \frac{3.91 \times 10^{-6} \mu_l^3 \, Re^{2.75}}{\eta_p d^3} A_a, \text{ where } Re = \frac{\rho u d}{\mu l}, \tag{6}$$

where $\mu_1 (kg\,m^{-1}s^{-1})$ (kg) is the dynamic viscosity of water, Re is the Reynolds number, $d$ (m) is the diameter of the pipes, u is the superficial velocity of flow (m/s), $\eta_p$ and $\eta_{elec}$ are the pump efficiency ($\eta_p = 0.7$), and $A_a$ (m²) is the tube aperture area. The pumps operate continuously.

## Artificial Lighting

Natural algae cultivation inherently revolves around the diurnal and seasonal cycles. To compensate for these cycles and to maximize the production of biomass, artificial lighting is used to allow 24-hour cultivation. Lights are turned on from dusk to dawn. Monthly averages of daylight hours are used to define the time the lighting system is in operation each day. The power ($P_a$) (W) consumed for artificially lighting the greenhouse area is calculated as [29]

$$P_a = A'(\frac{I_{avg}}{L_w C_f}).\qquad(7)$$

The intensity of the artificial lighting provided was set equal to the naturally available lighting in the month of July ($(I_{avg} = 1.7\,\mu E/m^2-s)$ over the entire greenhouse region ($A'=3345 m^2$). Specifications for high-efficiency fluorescent GRO lights [29] were used to estimate the power required for artificial lighting. The light intensity of the bulbs is expressed as $L_w = 220 Lu/W = 220\,Lu/W$ and the conversion factor ($C_f$) to convert between micromoles of photons (mE) and lux is 0.29.

## CO₂ Purification

Carbon dioxide acts as the only source of carbon for the biomass. Flue gases from power plants provide an inexhaustible source of $CO_2$. However, flue gases also contain varying levels of other gases such as $SO_X$ and $NO_X$ which are detrimental to microalgae culture beyond certain concentrations [30]. The monoethanolamine (MEA) absorption process can be used to separate pure $CO_2$ from flue gas for microalgae production. Kadam [31] determined that if about 18% of the total carbon dioxide consumed is taken directly from flue gases and the rest is purified through the MEA process, then, the toxic flue gases will be sufficiently low concentration for algae growth.

Molina Grima [19] determined that in order to make light the only limiting factor, $CO_2$ must be provided in excess and the ratio of the aqueous $CO_2$ concentration (kg/) to influent biomass concentration $C_i$ (kg/m$^3$) should be 0.63. Since growth rates for this system are lower than those in Molina's Grima study [19] due to reduced sunlight, this $CO_2$ represents a conservative estimate. The mass of carbon dioxide required was estimated based on this ratio, the media flow rate, and the influent biomass concentration. Although carbon dioxide has a high solubility in water, and it is likely that all $CO_2$ in the gas bubbled through the reactor would dissolve over the length of the reactor, a factor of safety of 2 was used here as an overestimate of the mass of $CO_2$ that would be required. The MEA $CO_2$ extraction process has been modeled and studied previously in context with algae production. Kadam [31, 32] reports that the process to extract $CO_2$ from flue gas and recover the MEA for reuse consumes 32.65 kWh per ton of $CO_2$ for algae cultivation. Details are not provided in these references to specifically quantify which of the steps in the MEA process consume the most electrical energy.

## Greenhouse Heating

Temperature control within the greenhouse is essential for algae cultivation in cold weather conditions. The energy consumed for greenhouse heating depends upon the total surface area exposed, insulation material, and temperature inside and outside the greenhouse. For a given greenhouse with surface area ($A_g$) (m$^2$) the heat loss per second ($Q_L$) (J/s) is given by [33]

$$Q_L = 1.05(\frac{1}{R})(T_{req} - T_{out})A_g. \tag{8}$$

where R (1.9 s) is the -value of the greenhouse insulating material; $T_{req}$ (25°C) and $T_{out}$ (°C) are the temperatures required within the greenhouse and outside the greenhouse, respectively. The greenhouse was assumed to be insulated with 10 mm twinwall polycarbonate with an -value of 1.9. The -value of insulated and windowless cultivation scenario was set at 30. The outside temperature $T_{out}$ is taken from monthly averages for Syracuse and Albany and is input as normal distributions for that month [34].

## Steam Drying and Dewatering

Algae are suspended in a dilute broth from photobioreactors [13]. Dewatering and drying of algae is necessary to reduce the water content to 5% [35] before the hexane oil extraction process. For algae with high vegetable oil content, it is suggested that continuous nozzle discharge centrifuges provide the best reliability and consume the least amount of energy. Centrifugation consumes 3.24 MJ/ of effluent media [36]. After centrifugation, algae water content is 70% (by weight).

Steam is utilized to further dry microalgae before oil extraction process. The natural gas consumed to provide the required steam energy was calculated based on the heat of vaporization of water, the mass of water that needed to be vaporized to reduce the water content from 0.70 to 0.05, and the efficiency of the boiler (0.93 [37]) and dryer (0.8 [26]) utilized. In the scenarios that utilize waste heat, it is assumed that because of the colocation of the algae production facility near a power plant, there is sufficient heat to dry the algae.

# Water Consumption

The consumption of water for the production of biofuels has recently been identified as a significant limitation to the development of an expanded biofuel economy. Water consumption occurs almost entirely in the feedstock production step for most biofuels. The average U.S. production biodiesel from soybeans requires 6,500 liters of water for evapotranspiration per liter of biodiesel produced [38]. Water consumption for algae biodiesel was calculated by a mass balance. The total water flowrate through the bioreactor is the sum of freshwater, water included in the algae recycle stream (35% recycle), and water recovered through the centrifuge dewatering process to increase the algae concentration from 0.34% to 30%. With this mass balance, 848 m$^3$ make up water is required annually or approximately 4 L water per L of biodiesel for the feedstock production stage. This represents approximately 99% of water recovery and reuse. In the transesterification and biodiesel cleaning processes, 1–3 L of water are required per L of biodiesel produced [39].

# Fertilizer Consumption

The microalgae culture media acts as the primary source of nutrients and carbon dioxide and a means of expelling excess oxygen. The minimum amount of nutrients consumed was defined based on the molecular formula of algae— [40]. N and P account for 6.5% and 1.3% of the algae mass. Assuming that maximum possible biomass concentration of algae cells is 4 kg / [13, 22] in a tubular PBR, the N and P consumed from the algae media would be $0.26 \, kg \, N/m^3$ and $0.052 \, kg \, P/m^3$. Excess fertilizer that passes through the bioreactor as part of the broth is assumed to be recovered in the centrifuge dewatering step for reuse. Since nearly all of the water is recycled, it is assumed that nearly all of the nutrients that are not consumed are also recycled.

# Utilizing GREET for Life Cycle Analysis

The GREET model was modified and used to calculate the energy use and emissions generated from algae production, oil extraction, and transesterification stages of biodiesel production as well as the upstream chemical and energy production processes. For a given fuel system GREET evaluates natural gas, coal, and petroleum use as well as the emissions of carbon dioxide equivalent greenhouse gases, volatile organic compounds, carbon monoxide, nitrogen oxides, particulates, and sulfur oxides from all lifecycle stages [14]. The GREET results are presented as primary energy consumed and emissions per million BTU fuel produced. The low heating value of the BD was used to convert to the functional unit used here—liters BD produced.

The GREET model is written in an MS Excel workbook and includes soy biodiesel production energy consumption and emissions pathways. A new spreadsheet page based on the soy biodiesel calculations was added to the GREET workbook and adapted for algae BD production. Default parameters for transesterification were used directly, but other input parameters including energy consumption for the various processes, biomass yield, nutrient requirements, carbon dioxide consumed were modified for algae biodiesel production based on the mass and energy flows presented above. The mix of electricity generation within New York State was used to define the primary energy consumed to generate electricity [41].

The extraction of oil from algae was assumed to be carried out by hexane oil extraction. The procedure is similar to soybean oil extraction, although significantly less hexane is required to recover oil from algae (0.030 kg of hexane/kg of dry algae) [11] than is required for soybeans (1.2 kg hexane/kg dried and flaked soy bean) [18]. During this process, algae meal is produced as a coproduct that can be used as an animal feed in the same manner that soy meal is used as a coproduct from soy biodiesel. GREET uses the displacement method to determine how much of the biomass production and extraction steps can defined as a credit for the biodiesel due to the production of a coproduct. The protein content of soy meal is 48% [42], as compared to 28% in algae meal [13] and 40% in soy beans [42]. Thus, 1 kg of algae meal displaces about 0.7 kg of soybean, whereas 1 kg of soy meal displaces about 1.2 kg of soy bean for animal feed. The credits for not having to produce 0.7 kg soy beans for every kg algae meal produced are subtracted from the total energy use and emissions associated with the algae production, oil extraction, and their associated upstream processes.

An additional credit was also attributed to the algae to represent the carbon dioxide sequestered from the power plant flue gas. Algae cell elemental composition was used to estimate the mass of carbon that was consumed by the algae growth within the PBR (0.51 kg of $CO_2$ consumed/kg algae grown).

# RESULTS

## Biomass Production

Biomass output is an important factor for determining life cycle energy analysis of microalgae biodiesel production. When natural lighting is used to minimize electricity consumption for artificial lighting, algae production rises steadily between the months of February and April (Figure 5). Biomass production is the highest between the months of May to July and is followed by a gradual decline in the months of

August to October. Production is the lowest in the winter months due to low natural irradiance. The uncertainty bars included represent 95% confidence intervals from Monte Carlo simulation outputs.

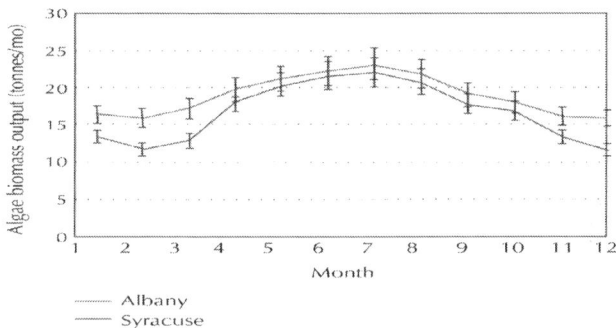

**Figure 5:** Algae Biomass production for Syracuse and Albany, NY, with natural lighting supplemented by artificial lighting for continuous algae production.

The annual biomass productivity in Albany is about 12% greater than that in Syracuse (Table4). These cities are at very similar latitudes, but the actual irradiance in Albany is higher due to less cloud cover. Biomass and subsequent biodiesel production in the windowless (artificial lighting only) scenario is much higher than greenhouse cases because illumination is maintained throughout the year at the highest level achieved naturally (noon in the month of July).

**Table 4:** Comparison of different locations and scenarios by biodiesel production

| Location | Biomass Produced | (tonnes/year) | Biodiesel produced ( $Lm^{-2}y^{-1}$ ) |
|---|---|---|---|
| Syracuse NY | Greenhouse Base Case | 202 | 19 |
| | Greenhouse w/waste heat | 202 | 19 |
| | Windowless Cultivation | 263 | 25 |
| Albany NY | Greenhouse Base Case | 225 | 21 |
| | Greenhouse w/waste heat | 225 | 21 |

# Energy Consumption for Microalgae Biodiesel Production

The energy consumed for biodiesel production was estimated by modeling individual processes in the algae cultivation stage. Energy required for the transesterification process is accounted directly by the GREET 1.8a model. The energy required for feedstock production through the drying process is illustrated in Figure 6. This does not include oil extraction and transesterification processes. Three variables can be assessed with this graph: location (Syracuse versus Albany), use of natural lighting versus solely artificial lighting and algae versus soybean production.

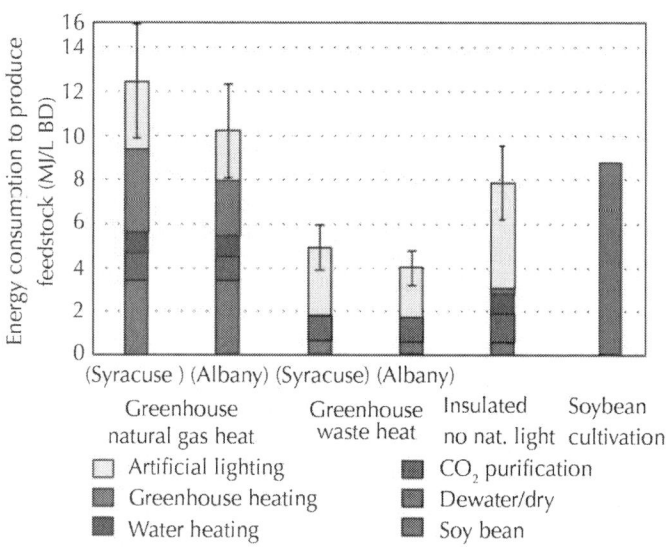

**Figure 6:** Energy consumption for microalgae and soy bean feedstock production. The error bars represent 95% confidence intervals on the total energy consumption for feedstock production.

Heating needs consume well over half of the total energy required for algae growth, dewatering, and drying. When no waste heat is available, dewatering and steam drying accounts for the greatest fraction—about 28–32% of the energy required for feedstock production. With the availability of waste heat, this component is reduced to about 13% of the total, which represents the electricity required for centrifugation. Greenhouse heating consumes a similar proportion of the total energy for algae production—about 25–30%. Water heating for cultivation consumes about 7–12% for feedstock production. Both locations have similar water heating requirements because groundwater temperature is assumed to be equal for both cases.

When natural lighting is utilized to the extent possible artificial lighting, it consumes about a quarter of the total energy required for algae cultivation. However, in the windowless cultivation case where there is no natural light available, the artificial lighting cost is almost doubled. However, the total energy requirements in this scenario are still less (35%) than the scenarios requiring natural gas to heat a greenhouse.

Among the design choices and trade-offs considered here, the growth and drying of algae with the utilization of waste heat is the only scenario that is substantially better than growing soybeans from the perspective of process energy consumed. These results clearly show the value of colocating an algae facility near a source of waste heat.

Overall, microalgae cultivation in Albany, NY, consumes about 18–21% less energy than Syracuse, NY, because greenhouse heating energy requirements are lower and higher natural lighting intensity yields about 12% higher biomass output.

Figure 7 illustrates the total lifecycle energy, which now also includes biodiesel production and credits for $CO_2$ consumption and algae/soy meal produced during the oil extraction phase. For most cases, the energy required for feedstock production is similar to the energy required for oil extraction and transesterification. Thus, the savings associated with the utilization of waste heat in the greenhouse also represent significant savings when the entire lifecycle energy consumption is considered. Greenhouse algae cultivation with waste heat in Albany consumes the least energy on a life cycle basis; however total energy consumption is very similar to that of the corresponding Syracuse case.

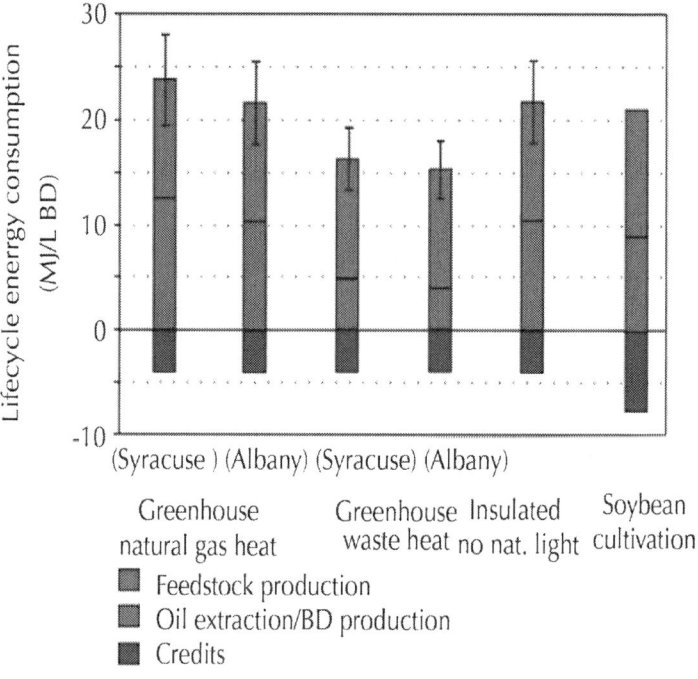

**Figure 7:** Total life cycle energy consumption by life cycle stage. The error bars represent 95% confidence intervals on the total lifecycle energy consumed.

The importance of the coproduct and carbon dioxide consumption credits are apparent from the data presented in Figure 7. Soy meal credits are higher than algae meal credits because of higher protein content and higher fraction of soy meal produced per liter of biodiesel (1 kg of algae meal displaces about 0.7 kg of soybean, whereas 1 kg of soy meal displaces about 1.2 kg of soy bean for animal feed). Adding the higher credits for the soy bean BD case to the energy required for production reduces the net energy for this case to a level below the well-insulated and windowless algae production scenario. The greenhouse scenarios utilizing waste heat are still the best option for minimizing the consumption of energy that has value for other uses.

Natural gas accounts for 65–80% of the total energy consumed on a life cycle basis for algae biodiesel production when waste heat is not available (data not shown). The high consumption of natural gas can be attributed to heating processes, the high fraction of natural gas in NY electricity mix (about 22%), and upstream consumption for process fuel and fertilizer production. In contrast, soy biodiesel requires substantially more petroleum (~5x) than microalgae consumes due to the extensive use of tractors and feedstock transportation when BD is made from soybeans. Thus, algae as a BD feedstock has a significant benefit over soybeans in terms of reducing our dependence on imported oil. Algae biodiesel production requires a significant amount of electricity and thus coal accounts for about 6–19% of the total life cycle energy consumption. Insulated cultivation has the highest coal consumption, about 19% of the total life cycle energy consumption, because of increased artificial lighting and electricity consumption. In comparison, for the greenhouse with waste heat case, only 7% the total lifecycle energy is derived from coal.

The processing of soybeans to prepare for oil extraction also requires some heating to dry the beans. Arguably, waste heat could be considered to reduce the fossil fuel consumption for soybean biodiesel too. However, whereas the algae feedstock could be grown at the same location where waste heat is available, the soybeans require a much more dispersed geographical region. Soybeans are typically transported 75 miles or less to a soybean crushing facility. Thus, the probability that soybean production and crushing facilities can be colocated with a waste heat source is significantly less than for algae. If this can be achieved, the lifecycle energy production for the feedstock production (green bar for soybean BD, Figure 7) would be less.

# Global Warming Potential

Global warming potential can be described as the impact of additional units of greenhouse gases to the atmosphere. The global warming potential for the different scenarios and gases is estimated in terms of carbon dioxide equivalents (Figure 8). All algae scenarios are allocated the same $CO_2$ credits because the carbon dioxide consumed per unit of algae produced is constant.

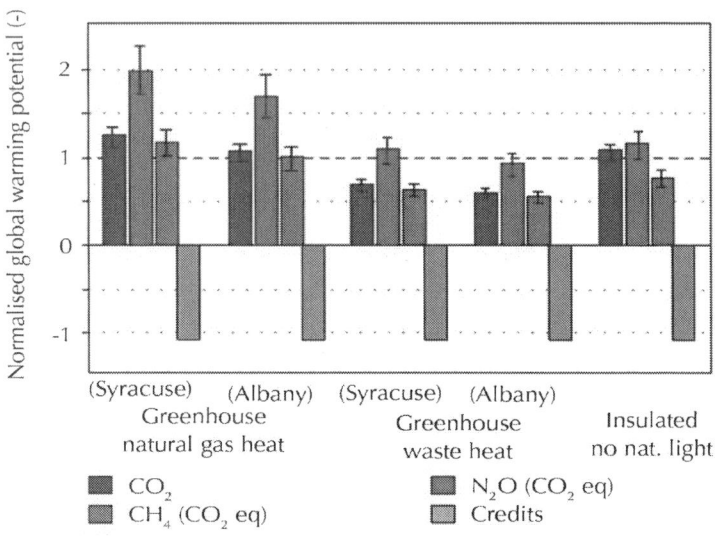

**Figure 8:** Global warming potential of microalgae biodiesel—mass emissions normalized by dividing by the corresponding emissions for soy biodiesel for comparison.

Most $CO_2$ emissions for algae biodiesel originate from upstream usage of energy use for heating, transportation fuel use, and coal combustion for electricity. The extraction and utilization of natural gas for heating use, electricity generation, and fertilizer production is accompanied by high methane emissions. Natural gas extraction has a very high methane emission factor. Overall, the emission of carbon dioxide is relatively low compared to methane due to the high natural gas use relative to petroleum or coal. Natural gas utilization has a much lower carbon dioxide emission factor than coal.

In cold climates, the production of algae biodiesel with the utilization of waste heat rather than natural gas consumption is the only approach that reduces greenhouse gas emissions relative to soy biodiesel.

## Other Air Emissions

The exposure of humans to air pollutants is increasingly associated with increased mortality and reduction in life expectancy [43]. Figure

9 presents the lifecycle air emissions for algae biodiesel production normalized to the corresponding air emissions estimated by GREET for soybean biodiesel. The microalgae biodiesel air emissions follow a trend similar to the total life cycle energy consumption. The high $NO_x$ emissions can be traced to high emission factors of equipment used to produce natural gas and the flaring of natural gas in refineries. The increased use of artificial lighting for the cultivation of algae in a windowless and well-insulated facility results in high particulate emissions, particularly in comparison to cases where natural lighting is used. These PM emissions originate mainly from coal and residual oil combustion use for electricity production.

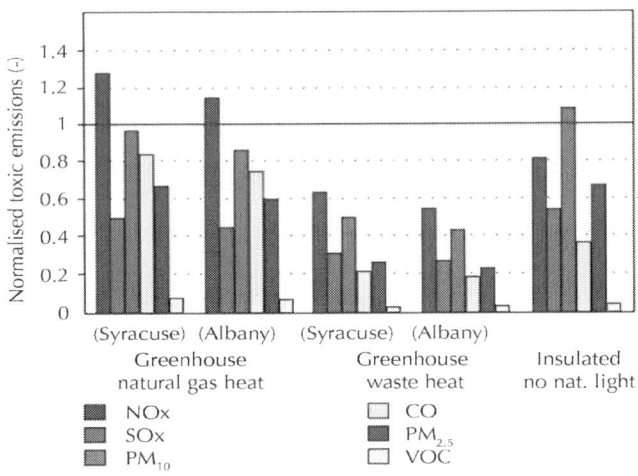

**Figure 9:** Toxic air emissions from microalgae biodiesel production—mass emissions normalized by dividing by the corresponding emissions for soy biodiesel for comparison.

VOC emissions from microalgae biodiesel production are much lower than soy biodiesel, because of low utilization of petroleum and hexane. The VOC emission factors for transportation fuels like gasoline are far greater than any other source. Thus, since algae is locally produced for biodiesel and transportation minimal, the VOC emissions from algae biodiesel are much less than soy biodiesel, primarily because only a minimal amount of hexane is required for extraction compared with soy beans.

Overall, the most important source of air emissions for microalgae is the upstream emissions associated with fuel and electricity generation. Yet, these emissions are still relatively low compared to soy biodiesel. The primary factor contributing to this apparent anomaly is the comparison of algae biodiesel produced in New York State to soy biodiesel produced nationally. NY State has a high percentage of hydroelectric (17%) and nuclear (29%) power production and relatively small amounts of electricity generated from coal (15%) [41]. this difference in upstream electricity generation has significant repercussions throughout the lifecycle energy emission estimates for any electricity-intensive manufacturing system: manufacturing in New York State benefits from relatively clean energy resources.

The acidification of soils and water bodies occurs mainly due to the transformation of gaseous pollutants ($SO_x$,$NO_x$) into acids. The acidification potential of the different cases is estimated in $SO_2$ equivalents. All cases of microalgae biodiesel are better than soy biodiesel in terms of acidification emissions. The total $SO_2$ equivalents follow a trend that resembles the total energy usage.

# Summary of Results

A summary of the lifecycle sustainability assessment metrics for the various algae biodiesel production scenarios and soy biodiesel production is presented in Table 5. The most sustainable biodiesel production for all cases requires the colocation of the algae and BD production facility in the vicinity of a source of waste heat. "Free" heat greatly reduces the fossil fuel consumption and all related greenhouse gas and other air pollutants. At a similar latitude, choosing a location that maximizes sunlight helps somewhat to increase the algae production rate and, therefore, reduce the impacts when the results are compared on a per BD-produced basis. These effects are small, however, compared to the benefits of utilizing waste heat. Similarly, a well-insulated facility can help reduce heating needs, but the consequences of increased electricity use for artificial lighting decrease the benefits of reduced heating fuel required. In most regions of the U.S., where a higher fraction of the electricity mix is generated from fossil fuels, the well-insulated windowless scenario would be worse in terms of most sustainability metrics due to the increased dependence on fossil fuels.

**Table 5:** Summary of average sustainability metrics to compare algae and soy BD production

| Environmental Impact | | Greenhouse Nat. Gas, Syracuse | Greenhouse Nat. Gas, Albany | Greenhouse w/waste heat, Syracuse | Greenhouse w/waste heat, Albany | Insulated, no nat. light | Soy biodiesel production |
|---|---|---|---|---|---|---|---|
| Total life cycle energy Consumption (MJ/L of BD) | | 23 | 21 | 16 | 15 | 22 | 21 |
| Land utilization (m2/L of BD/yr) | | 0.053 | 0.048 | 0.053 | 0.048 | 0.040 | 22.2 |
| Water Consumption (L water/L BD) | | 5-7 | 5-7 | 5-7 | 5-7 | 4-6 | 6,500 |
| Greenhouse gas emissions (g $CO_2$ equiv/L of BD) | | 1350 | 1150 | 740 | 630 | 910 | 925 |
| cidification potential (g $SO_2$ eq./L of BD) | | 4.9 | 4.6 | 2.8 | 2.5 | 3.4 | 4.0 |
| Toxic Emissions (g /L of BD) | PM 10 | 5.1 | 4.6 | 2.6 | 2.3 | 5.7 | 5.3 |
| | PM 2.5 | 1.8 | 1.6 | 0.7 | 0.6 | 1.8 | 2.7 |
| | VOC | 0.22 | 0.20 | 0.06 | 0.05 | 0.09 | 3.4 |
| | CO | 2.4 | 2.1 | 0.6 | 0.5 | 1.0 | 2.8 |

*does not include credits.

# CONCLUSIONS

Cultivation of microalgae in NY State is an energy intensive process owing to temperature control and steam drying process. Colocating microalgae cultivation with a power plant is highly desirable. Annual production of microalgae requires the utilization of waste heat for steam drying, water heating, and greenhouse heating in order to be substantially better than soy biodiesel in terms of energy consumption and emissions. When waste heat is utilized, microalgae biodiesel production consumes less energy than soy biodiesel.

Microalgae consumes less than one third the petroleum fossil fuel required for soy biodiesel and only a small fraction of the water. The feasibility of microalgae biodiesel production at a given location is greatly dependent on availability of waste heat and natural lighting conditions. The availability of either one or both makes algae biodiesel production process cleaner in terms of air emissions and consumes much less energy than soy biodiesel. However if both natural lighting and waste heat are absent, algae biodiesel production consumes more energy than soy biodiesel production and emits an equal or more amount toxic air emissions.

Coproducts produced during algae biodiesel production process have less protein content than soy meal and, thus, are less valuable. The production of high value coproducts allows for increased energy allocation for soy biodiesel and thus emissions or energy consumption of both the feedstocks is very close and comparable.

Most microalgae biodiesel production scenarios have low or very similar emissions as compared to soy biodiesel. Greenhouse gas emissions for algae biodiesel are generally higher than soy biodiesel except when waste heat is utilized, in which case emissions are equal. The emission of volatile organic compounds for soy biodiesel is much higher than that for algae biodiesel. Emissions from microalgae production originate mainly from upstream fossil fuel energy consumption. Reducing needs for unit processes like greenhouse heating, lighting, and other systems will have significant benefits.

# APPENDIX

## Estimation of Average Light Intensity

A model is used to estimate the growth rate as a function of light for a geographic location and day of the year for tubular photobioreactors [12] that takes into account various solar angles, cloudiness, and reactor geometry. The model accounts for weather conditions and relates them to the growth of biomass.

The photosynthetically active irradiance (I) ($\mu Em^{-2}s^{-1}$) used in (2) is a function of various solar angles (w,$w_s$)total solar irradiance ($Jm^{-2}d^{-1}$), and photosynthetic efficiency ($E_f$) ($EJ^{-1}$). The total photosynthetically active irradiance (I) over a horizontal culture surface can be determined from the total solar irradiance (H) ($kWhm^{-2}d^{-1}$) directly by utilizing the following equation [22, 44]:

$$I = \frac{\pi HE}{24}(a + b\cos w)\,|\,(\frac{\cos w(\cos w_s)}{\sin w_s - w_s(\cos w_s)})\,|, \qquad \text{(A.1a)}$$

Where

$$a = 0.409 + 0.502\sin(w_s - 60), \qquad \text{(A.1b)}$$

$$b = 0.661 + 0.477\sin(w_s - 60). \qquad \text{(A.1c)}$$

The photosynthetic efficiency ($E_f$) is the ratio of the photosynthetically active radiation to the total incident radiation (H). $E_f$ for P. tricornutum algae varies as (1.74 $\pm$ 0.09 *$10^{-6}$ $EJ^{-1}$ [22]. It is substituted as a normal distribution with mean 1.74 *$10^{-6}$ and standard deviation 0.09 *$10^{-6}$ for Monte Carlo simulations.

The total extraterrestrial irradiance incident on the earth ($H_0$) (kWh $m^{-2}$ $d^{-1}$) varies seasonally. Some of the incoming solar irradiance is lost due to varying atmospheric transmissivity associated with cloud cover. The atmospheric transmissivity, which is also known as the clearness index, is the ratio of the total daily radiation (H) (kWhm$^{-2}$ d$^{-1}$) at ground level to the total daily extraterrestrial radiation ($H_0$) (kWh m$^{-2}$ d$^{-1}$). , a unitless parameter, varies by month. For this study, NREL's solar data [24] were used as monthly averages for $H_0$ and $K_h$ values [24]. The total irradiance incident that makes it to the earth surface (H) surface was calculated as the product of $H_0$ and .$K_h$ Values for H in Syracuse NY are shown in Figure 10.

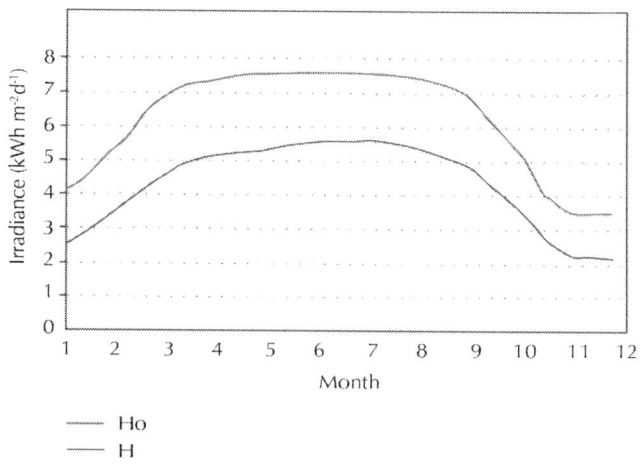

**Figure 10:** Solar irradiance for Syracuse, NY.

Seasonal variations in solar irradiance may be attributed to changes in the solar declination angle. The solar declination angle varies throughout the year and is a function of the day number of the year (N). N takes on values from 1 (January 1) to 365. The solar declination angle δ is constant for all locations in the northern hemisphere and can be calculated for each day as follows [45]

$$\delta = 23.45\sin(\frac{360}{365}(284+N)). \qquad (A.2)$$

Note

23.45$^0$ is the angle at which the axis of the earth is tilted and 360$^0$ and 284 are conversion factors from radians to degree. 365 denotes the number of days taken to complete one revolution by the earth around the sun.

The solar angle at sunrise ($w_s$) incorporates seasonal variations in solar declination angle with the latitude ($\phi$) (degrees) of the area being considered:

$$\cos w_s = -(\tan \delta)(\tan \Phi). \qquad (A.3)$$

The solar hour angle (w) (degrees) for a location on earth is zero when the sun is directly overhead and negative before noon and positive in the afternoon [46]. Similarly, hourly changes in the solar irradiance depend upon solar hour angle (w), which is a function of the solar hour (h). The solar hour h varies from 24 to 1 [43]:

$$w = 15(12 - h). \qquad (A.4)$$

Equations (A.2)–(A.4) can be used in conjunction with (A.1a)–(A.1c) to find photosynthetically active irradiance at any given latitude and time.

# ACKNOWLEDGMENTS

This work was supported by United States Department of Agriculture Grant no. NRCS-68-3A75-6-512. Publication of this work does not imply endorsement by the funding agency.

# REFERENCES

1.    DOE Alternative Fuels and Advanced Vehicles Data Center, Biodiesel Production. October, 2009, http://www.afdc.energy.gov/afdc/fuels/biodiesel_production.html.
2.    D. Pimentel, S. Williamson, C. E. Alexander, O. Gonzalez-Pagan, C. Kontak, and S. E. Mulkey, "Reducing energy inputs in the US food system," Human Ecology, vol. 34, no. 4, pp. 459–471, 2008. View at Publisher · View at Google Scholar · View at Scopus

3.  C. U. Ugwu, H. Aoyagi, and H. Uchiyama, "Photobioreactors for mass cultivation of algae," Bioresource Technology, vol. 99, no. 10, pp. 4021–4028, 2008. View at Publisher ·View at Google Scholar · View at Scopus

4.  A. Banerjee, R. Sharma, Y. Chisti, and U. C. Banerjee, "Botryococcus braunii: a renewable source of hydrocarbons and other chemicals," Critical Reviews in Biotechnology, vol. 22, no. 3, pp. 245–279, 2002. View at Publisher · View at Google Scholar · View at Scopus

5.  Y. Li, M. Horsman, N. Wu, C. Q. Lan, and N. Dubois-Calero, "Biofuels from microalgae," Biotechnology Progress, vol. 24, no. 4, pp. 815–820, 2008. View at Publisher· View at Google Scholar · View at Scopus

6.  E. Ono and J. L. Cuello, "Feasibility assessment of microalgal carbon dioxide sequestration technology with photobioreactor and solar collector," Biosystems Engineering, vol. 95, no. 4, pp. 597–606, 2006. View at Publisher · View at Google Scholar · View at Scopus

7.  M. Olaizola, "Commercial development of microalgal biotechnology, from the test tube to the marketplace," Biomolecular Engineering, vol. 20, no. 4–6, pp. 459–466, 2003. View at Publisher · View at Google Scholar · View at Scopus

8.  A. Richmond, "Microalgal biotechnology at the turn of the millenniumml: a personal view," Journal of Applied Phycology, vol. 12, no. 3–5, pp. 441–451, 2000. View at Scopus

9.  S. Hirata, M. Hayashitani, M. Taya, and S. Tone, "Carbon dioxide fixation in batch culture of Chlorella sp. using a photobioreactor with a sunlight-collection device,"Journal of Fermentation and Bioengineering, vol. 81, no. 5, pp. 470–472, 1996. View at Publisher · View at Google Scholar · View at Scopus

10. J. Sheehan, T. Dunahay, J. Benemannm, and P. Roessler, "A look back at the U.S. Department of Energy›s Aquatic Species Program-Biodiesel from Algae," Tech. Rep. TP-580-24190, NREL, Golden, Colo, USA, 1998,http://www.fuelandfiber.com/Athena/biodiesel_from_algae_es.pdf.

11. H. Xu, X. Miao, and Q. Wu, "High quality biodiesel production from a microalga Chlorella protothecoides by heterotrophic growth in fermenters," Journal of Biotechnology, vol. 126, no. 4,

pp. 499–507, 2006. View at Publisher · View at Google Scholar · View at Scopus

12. E. Molina, J. Fernandez, F. G. Acién, and Y. Chisti, "Tubular photobioreactor design for algal cultures," Journal of Biotechnology, vol. 92, no. 2, pp. 113–131, 2001. View at Publisher · View at Google Scholar · View at Scopus

13. Y. Chisti, "Biodiesel from microalgae," Biotechnology Advances, vol. 25, no. 3, pp. 294–306, 2007. View at Publisher · View at Google Scholar · View at Scopus

14. M. Wang, "Overview of GREET model development at Argonne," in Proceedings of the GREET User Workshop, Center for Transportation Research, Argonne National Laboratory, Sacramento, Calif, USA, March 2008, http://www.transportation.anl.gov/pdfs/TA/468.pdf.

15. A. Lavigne and S. E. Powers, "Evaluating fuel ethanol feedstocks from energy policy perspectives: a comparative energy assessment of corn and corn stover," Energy Policy, vol. 35, no. 11, pp. 5918–5930, 2007. View at Publisher · View at Google Scholar · View at Scopus

16. J. Hill, S. Polasky, E. Nelson, et al., "Climate change and health costs of air emissions from biofuels and gasoline," Proceedings of the National Academy of Sciences of the United States of America, vol. 106, no. 6, pp. 2077–2082, 2009. View at Publisher · View at Google Scholar · View at Scopus

17. CARB, "Detailed California-GREET Pathway for Biodiesel (Esterified Soyoil) from Midwest Soybeans, Ver. 2.1," Tech. Rep., California Air Resources Board Stationary Source Division, 2009, http://www.arb.ca.gov/fuels/lcfs/022709lcfs_biodiesel.pdf.

18. J. Sheehan, V. Camobreco, J. Duffield, M. Graboski, and H. Shapouri, "Life cycle inventory of biodiesel and petroleum diesel for use in an urban bus," Tech. Rep. SR-580-24089, NREL, Golden, Colo, USA, 1998.

19. E. Molina Grima, E.-H. Belarbi, F. G. A. Fernandez, A. R. Medina, and Y. Chisti, "Recovery of microalgal biomass and metabolites: process options and economics," Biotechnology Advances, vol. 20, no. 7-8, pp. 491–515, 2003. View at Publisher · View at Google Scholar · View at Scopus

20. G. L. Rorrer and R. K. Mullikin, "Modeling and simulation of a tubular recycle photobioreactor for macroalgal cell suspension cultures," Chemical Engineering Science, vol. 54, no. 15-16, pp. 3153–3162, 1999. View at Publisher · View at Google Scholar ·View at Scopus

21. M. Barbosa, Microalgal photobioreactors: scale up and optimization, Ph.D. Dissertation, van Wageningen University, Wageningen, The Netherlands, 2003.

22. E. Molina Grima, F. G. A. Fernandez, F. G. Camacho, and Y. Chisti, "Photobioreactors: light regime, mass transfer, and scaleup," Journal of Biotechnology, vol. 70, no. 1–3, pp. 231–247, 1999. View at Publisher · View at Google Scholar · View at Scopus

23. F. G. A. Fernandez, F. G. Camacho, J. A. S. Perez, J. M. F. Sevilla, and E. M. Grima, "Modeling of biomass productivity in tubular photobioreactors for microalgal cultures: affects of dilution rate, tube diameter, and solar irradiance," Biotechnology and Bioengineering, vol. 58, no. 6, pp. 605–616, 1998. View at Publisher · View at Google Scholar · View at Scopus

24. National Renewable Energy Laboratory, "National solar radiation database 1991–2005 update: user›s manual," Tech. Rep. TP-581-41364, NREL, Golden, Colo, USA, 2007.

25. L. A. Meireles, A. C. Guedes, C. R. Barbosa, J. L. Azevedo, J. P. Cunha, and F. X. Malcata, "On-line control of light intensity in a microalgal bioreactor using a novel automatic system," Enzyme and Microbial Technology, vol. 42, no. 7, pp. 554–559, 2008. View at Publisher · View at Google Scholar · View at Scopus

26. American Council for Energy Efficient Economy. Water Heating, January 2009,http://www.aceee.org/consumerguide/waterheating.htm.

27. Noritz America Corporation. US Average Groundwater Temperature, November, 2008,http://www.noritz.com/u/US_ground_temperature%5B1%5D.pdf.

28. G. Burgess, J. G. Fernandez-Velasco, and K. Lovegrove, "Materials, geometry, and net energy ratio of tubular photobioreactors for microalgal hydrogen production," inProceedings of the World Hydrogen Energy Conference (WHEC ‹06), vol. 16, Lyon, France, June 2006.

29. OSRAM Sylvania Corp, "Photosynthetically Active Radiation Units," October, 2009,http://openwetware.org/images/e/e8/ Conversion_lux.pdf.

30. M. H. Reddy, Application of algal culture technology for carbon dioxide and flue gas emission control, Master of Science Thesis, Arizona State University, 2002,http://www4.eas.asu.edu/pwest/ Theses_Diss/Madhu_Thesis%20Algae%20Photosynthesis.pdf.

31. K. L. Kadam, "Microalgae production from power plant flue gas: environmental implications on a life cycle basis," Tech. Rep. TP-510-29417, National Renewable Energy Laboratory, Golden, Colo, USA, 2001.

32. K. L. Kadam, "Power plant flue gas as a source of CO2 for microalgae cultivation: economic impact of different process options," Energy Conversion and Management, vol. 38, supplement 1, pp. S505–S510, 1997. View at Scopus

33. ACF Greenhouses, September, 2008, http://www.littlegreenhouse. com/heat-calc.shtml.

34. Weather.com, Syracuse monthly average temperatures, November, 2008,http://www.weather.com/outlook/driving/ interstate/wxclimatology/monthly/graph/13201?from=month_ bottomnav_driving.

35. J. M. Lidell, "Extraction of triglycerides from microorganisms," US Patent no. 6180376, 2001, http://www.patentstorm.us/ patents/6180376/description.html.

36. E. M. Grima, E.-H. Belarbi, F. G. A. Fernandez, A. R. Medina, and Y. Chisti, "Recovery of microalgal biomass and metabolites: process options and economics," Biotechnology Advances, vol. 20, no. 7-8, pp. 491–515, 2003. View at Publisher · View at Google Scholar· View at Scopus

37. S. Sanford, "Greenhouse unit heaters: types, placement, and efficiency," 2008,http://learningstore.uwex.edu/pdf/A3784-15. pdf.

38. R. Dominguez-Faus, S. E. Powers, J. G. Burken, and P. J. Alvarez, "The water footprint of biofuels: a drink or drive issue?" Environmental Science and Technology, vol. 43, no. 9, pp. 3005–3010, 2009. View at Publisher · View at Google Scholar · View at Scopus

39. National Research Council, Water Implications of Biofuels Production in the United States, National Academies Press, Washington, DC, USA, 2008.

40. J. U. Grobbelaar, "Algal nutrition," in Handbook of Microalgal Culture: Biotechnology and Applied Phycology, A. Richmond, Ed., Blackwell Publishing, Israel, 2004.

41. DOE Energy Information Agency, New York Electricity Profile, 2007,http://www.eia.doe.gov/cneaf/electricity/st_profiles/new_york.html.

42. H. Huo, M. Wang, C. Bloyd, and V. Putsche, "Life-cycle assessment of energy and greenhouse gas effects of soybean-derived biodiesel and renewable fuels," Environmental Science and Technology, vol. 43, no. 3, pp. 750–754, 2008.

43. Environmental Protection Agency. Health and Environmental Effects of Particulate Matter, December, 2008, http://www.epa.gov/ttncaaa1/naaqsfin/pmhealth.html.

44. J. A. Duffie and W. A. Beckman, Solar Engineering of Thermal Processes, Wiley, New York, NY, USA, 1980.

45. B. Y. H. Liu and R. C. Jordan, "The interrelationship and characteristic distribution of direct, diffuse and total solar radiation," Solar Energy, vol. 4, no. 3, pp. 1–19, 1960. View at Scopus

46. Sunlit Design, Solar Hour Angle, October, 2008, http://www.sunlit-design.com/infosearch/hourangle.php.

Chapter 5

# Utilization of Biodiesel By-Products for Biogas Production

Nina Kolesárová, Miroslav Hutňan, Igor Bodík, and Viera Špalková

Institute of Chemical and Environmental Engineering, Faculty of Chemical and Food Technology, Slovak University of Technology, Radlinského 9, 812 37 Bratislava, Slovakia

Received 22 August 2010; Revised 22 November 2010; Accepted 7 January 2011

## ABSTRACT

This contribution reviews the possibility of using the by-products from biodiesel production as substrates for anaerobic digestion and production of biogas. The process of biodiesel production is predominantly carried out by catalyzed transesterification. Besides desired methylesters, this reaction provides also few other products, including crude glycerol, oil-pressed cakes, and washing water. Crude glycerol or g-phase is heavier separate liquid phase, composed mainly by glycerol. A couple of studies have demonstrated the possibility of biogas production, using g-phase as a single substrate, and it has also shown a great potential as a cosubstrate by anaerobic treatment of different types of organic waste or energy crops. Oil cakes or oil meals are solid residues obtained after oil extraction from the seeds. Another possible by-product is the washing water from raw biodiesel purification, which is an oily and soapy liquid. All of these materials have been suggested as feasible substrates for anaerobic degradation, although some issues and inhibitory factors have to be considered.

# INTRODUCTION

Renewable energy sources and biofuels, including biodiesel, have been gaining increasing attention recently as a replacement for fossil fuels [1]. However, their implementation in the general market depends on making these fuels more competitive. A convenient way to lower the costs of biofuels is to use the by-products as a potential source of energy, rather then treat them as waste.

Biodiesel is a prominent candidate as alternative diesel fuel. It is offering few advantages compared to conventional diesel, including the status of renewable energy source and lower emissions. Advances against petroleum diesel fuel are represented by the terms of sulfur content, flash point, content of aromatic substances, and biodegradability [1].

With approximately 245 processing plants and annual production of about 9 million tons, European Union has had the leading position in both production and consumption of biodiesel [2]. These plants are mainly located in Germany, Italy, Austria, France, and Sweden. Production of biodiesel has been expanding rapidly also on the other continents, mainly in the USA and developing countries, such as India, Brazil, Argentina, Malaysia, and Fiji.

As a primary feedstock, vegetable oils, animal fats, or waste cooking oils can be used for the production of biodiesel. In Europe, rapeseed oil is predominantly used, while, in the world extent, highest quantities of biodiesel are produced from soya oil [3].

The process of biodiesel production is usually carried out by catalyzed transesterification with alcohol, most likely methanol (Figure 1). A catalyst is usually involved to improve the reaction rate and yield [3]. Alkalies (sodium hydroxide, potassium hydroxide, carbonates, and corresponding sodium and potassium alkoxides), acids (sulfuric acid, sulfonic acid or hydrochloric acid), or enzymes can be used to catalyze the reaction. Base-catalyzed transesterification is much faster than the acid-catalyzed one (base catalyzed transesterif ication is basically finished within one hour) and is most often used commercially [4–6].

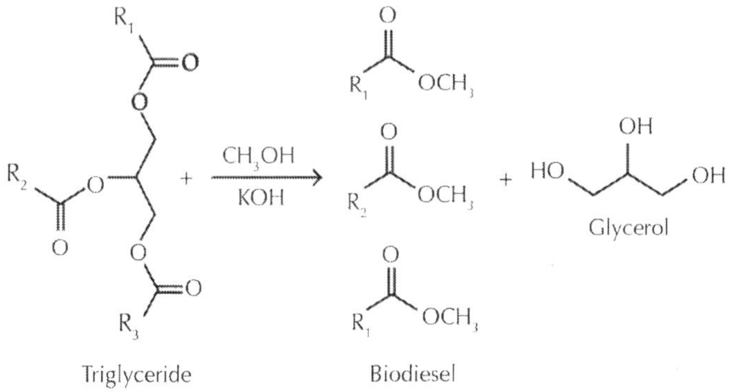

**Figure 1**: Reaction of biodiesel production by base-catalyzed transesterification with methanol.

Besides the desired methylesters this reaction provides also few other products. Isolation of oil from the oil seed plants by pressing and extraction provides oil cakes or oil meal as a by-product. In the reaction of transesterification, triglycerides are converted into glycerol and methylesters (biodiesel) are separated from the heavier glycerol phase (crude glycerine) by settling. To remove the impurities, soaps, short chain fatty acids and excess methanol, crude biodiesel is subsequently washed generating the washing water as another potential by-product.

Although the present production of biodiesel is mainly carried out by homogeneous base-catalyzed transesterification, implementation of alternative approaches has been increasingly studied [7–24].

Homogeneous base catalysis has few disadvantages, such as the high requirement for the purity of oil. The high consumption of energy and costly separation of the homogeneous catalyst from the reaction mixture have also called for development of new catalysts.

Different alternative techniques have been applied for biodiesel production [7–11]. The most promising ones include employing heterogeneous catalyst, lipase catalyst or supercritical alcohol [12].

Catalysis using solid heterogeneous catalysts runs slower than homogeneouscatalysis, however it can be integrated with continuous processing technologies. A great variety of catalysts in catalytic transesterification of vegetable oils have been used recently, including zeolites, hydrotalcites, oxides, and so forth, [13–17]. Utilization of heterogeneous catalysts provides few advantages over the homogeneous base catalysis, including mainly the easier purification of methylesters from glycerol and impurities. Also, the content of free fatty acids and water in the raw material does not affect the reaction [12].

Enzyme catalyzed transesterification using lipases for biodiesel production is also increasingly studied [18–20]. Similarly to heterogeneous catalysis, it also provides a solution of avoiding difficult recovery of glycerol and methylesters purification. Although this technology offers an attractive alternative, the industrial application has been slow due to feasibility aspects and some technical challenges, resulting from the low solubility of methanol and glycerol in biodiesel and high cost of lipases as catalyst.

Transesterification with supercritical methanol provides several advantages, compared to traditional methods [21–24]. The reaction is fast, in addition no catalyst is needed and therefore the separating process of the catalyst and saponified products becomes unnecessary. Generation of washing water can also be avoided. However the high pressure and temperature (239–385°C) is required, which leads to high energy consumption and production costs.

Commonly the most important by-products from biodiesel production are pressed cakes from oil extraction, crude glycerol and washing water [5, 7, 18]. The nature of these products is highly dependent on the character of raw material and processing technique, although generally they present suitable substrates for anaerobic digestion with the production of biogas.

# CRUDE GLYCEROL

Crude glycerol (g-phase) is heavier separate liquid phase, composed mainly by glycerol. In general for every 100 kilograms of biodiesel about 10 kilograms of g-phase is produced.

Crude glycerol generated by homogeneous base-catalyzed transesterification contains approximately 50–60% of glycerol, 12–16% of alkalies especially in the form of alkali soaps and hydroxides, 15–18% of methyl esters, 8–12% of methanol, 2-3% of water and further components [25, 26]. Tables 1 and 2 summarize the characteristics of g-phases based on the source of oil used for the production of biodiesel. Analytical results from the macroelement screening tests are listed in Table 1. Crude glycerol contains a variety of elements, such as calcium, magnesium, phosphorus or sulfur, originating from the primary oil. Larger quantities of sodium or potassium are also contained, coming from the catalyst.

**Table 1**: Analysis results of macroelements, carbon and nitrogen in crude glycerol from different feedstocks (BDL indicates values that are below the detection limit for the corresponding analytical method) [25]

| Feed stocks | Ida Gold Mustard | Pac Gold Mustard | Rapeseed | Canola | Soybean | Crambe | Waste vegetable oils |
|---|---|---|---|---|---|---|---|
| Calcium, ppm | 11.7 ± 2.9 | 23.0 ± 1.0 | 24.0 ± 1.7 | 19.7 ± 1.5 | 11.0 ± 0 | 163.3 ± 11.6 | BDL |
| Potassium, ppm | BDL | BDL | BDL | BDL | BDL | 216.7 ± 15.3 | BDL |
| Magnesium, ppm | 3.9 ± 1.0 | 6.6 ± 0.4 | 4 ± 0.3 | 5.4 ± 0.4 | 6.7 ± 0.2 | 126.7 ± 5.8 | 0.4 ± 0 |
| Phosphorus, ppm | 25.3 ± 1.2 | 48 ± 2.0 | 65 ± 2.0 | 58.7 ± 6.8 | 53.0 ± 4.6 | 136.7 ± 57.7 | 12.0 ± 1.5 |
| Sulfur, ppm | 21.0 ± 2.9 | 16.0 ± 1.4 | 21.0 ± 1.0 | 14.0 ± 1.5 | BDL | 128.0 ± 7.6 | 19.0 ± 1.8 |
| Sodium, % wt | 1.17 ± 0.15 | 1.23 ± 0.12 | 1.06 ± 0.07 | 1.07 ± 0.12 | 1.2 ± 0.1 | 1.10 ± 0.10 | 1.40 ± 0.16 |
| Carbon, % wt | 24.0 ± 0.00 | 24.3 ± 0.58 | 25.3 ± 0.58 | 26.3 ± 0.58 | 26.0 ± 1 | 24.0 ± 0.00 | 37.7 ± 0.58 |
| Nitrogen, % wt | 0.04 ± 0.02 | 0.04 ± 0.01 | 0.05 ± 0.01 | 0.05 ± 0.01 | 0.04 ± 0.03 | 0.06 ± 0.02 | 0.12 ± 0.01 |

**Table 2:** Food nutrient analysis for crude glycerol samples [25]

| Feed stocks | Ida Gold Mustard | Pac Gold Mustard | Rapeseed | Canola | Soybean | Crambe | Waste vegetable oils |
|---|---|---|---|---|---|---|---|
| Fats, % | 2.03 | 1.11 | 9.74 | 13.1 | 7.98 | 8.08 | 60.1 |
| Carbohydrates, % | 82.8 | 83.8 | 75.5 | 75.2 | 76.2 | 78.6 | 26.9 |
| Protein, % | 0.14 | 0.18 | 0.07 | 0.06 | 0.05 | 0.44 | 0.23 |
| Calories, kJ/kg | 14.6 | 14.5 | 16.3 | 17.5 | 15.8 | 16.3 | 27.2 |
| Ash, % | 2.8 | 1.9 | 0.7 | 0.65 | 2.73 | 0.25 | 5.5 |

Table 2 shows the content of protein, fat, ash, carbohydrates in percents and caloric value for kg. G-phase is mostly composed of carbohydrates, represented by glycerol. The ash contained in crude glycerol is mainly sodium or potassium from the catalyst.

Considering that the processing technology of biodiesel production affects the characteristics of by-products, the new technologies and modern catalysts can be expected to influence the composition and utilization of crude glycerol. For example, g-phase originating from biodiesel production using rapeseed oil with heterogeneous catalyst is limpid and colorless, containing at least 98% of glycerin and neither ash, nor inorganic compounds were detected in it [27].

As the biodiesel production is increasing exponentially, the crude glycerol generated in this process has also been generated in a large quantity. Despite the wide applications of pure glycerol in pharmaceutical, food and cosmetic industries, the refining of crude glycerol to a high purity is too expensive, especially for small and medium biodiesel producers [28]. The investments for the construction and startup operation of crude glycerol purification facility make according to Singhabhandhu [29] roughly 65 million Euros (facilities with production capacities of 1.4–2 ML/y). Weber [30] mentions 27% of the capital investment costs going to construction of the technical glycerin facility in a 12 ML biodiesel refinery (Aschach, Austria).

To improve the economic feasibility of biodiesel industry, new alternate ways of utilization of g-phase have been studied recently. Possibilities such as combustion, coburning, composting, animal feeding, thermochemical conversions and biological conversion have been applied for crude glycerol processing [31–46].

One of the possible applications is utilization of g-phase as carbon and energy source for microbial growth in industrial microbiology. Microbial conversion of glycerol to various compounds has been investigated recently, with particular focus on the production of 1,3-propanediol [47–50], which has been considered as a main product of glycerol fermentation [28]. 1,3-propanediol presents several interesting applications, it can be used as a monomer for polycondensations to produce plastics with special properties, (polyesters, polyethers and polyurethanes) [51–54] as a monomer for cyclic compounds, as a polyglycol-type lubricant [55] and it also may serve as a solvent [56]. The biotechnological production of 1,3-propanediol from glycerol has been demonstrated for several bacteria, Klebsiella pneumoniae, Clostridium butyricum and Citrobakter freundii have been most commonly used in the studies [47–50, 57–59].

Besides the production of 1,3-propanediol, glycerol can also be used as a carbon source to obtain other valuable microbial products, such as recombinant proteins and enzymes, microbial lipids (single-cell oils), medicinal drugs, antibiotics and fine chemicals. Bioconversion of g-phase into chemicals, such as dihydroxyacetone, 1,2-propanediol, ethanol, hydrogen,citric acid, propionic acid, polyglycerols, succinate, have been also increasingly studied recently [60–77].

Another option offers biological production of methane from crude glycerol using anaerobic sludge [78–88]. Besides the production of methane, the advantages include low nutrient requirements, energy savings, generation of low quantities of sludge and excellent waste stabilization. Glycerol is a readily digestible substance, which can be easily stored over a long period. High energy content in g-phase makes it an interesting substrate for anaerobic digestion as well, since it offers high production of biogas in smaller reactor volumes. A great variety of microorganisms is able to use this substrate as a carbon source for the growth under anaerobic conditions, such as Citrobacter freundii, Klebsiella pneumoniae, Clostridium pasteurianum, Clostridium butyricum, Enterobacter agglomerans, Enterobacter aerogenes or Lactobacillus reuteri [28, 34, 58, and 67]. The production of biogas through anaerobic digestion offers significant advantages over other forms of crude glycerol treatment. It requires lower investments and simpler operational conditions compared to more sophisticated preprocessing technologies, which makes it ideal for local applications. Less biomass sludge is produced in comparison to aerobic treatment technologies. The digestate is an improved fertilizer in terms of both its availability to plants and its rheology. A source of carbon neutral energy is produced in the form of biogas.

# ANAEROBIC DIGESTION OF CRUDE GLYCEROL

Considering anaerobic treatment of crude glycerol, potential of its main component glycerol has been well-known for a longer period

[89–91]. Digestion of pure glycerol has been investigated both as a primary substrate [89, 90], and as an intermediate product of anaerobic degradation of fats [91]. Biodegradation have been carried out using either pure cultures of microorganisms [90] or sludge composed of mixed cultures from wastewater treatment plant [89].

Few studies focused on biogas production from g-phase [78–88] have also been realized recently. Anaerobic treatment of g-phase as a single substrate [78–81] was carried out as well as coprocessing of crude glycerol with different substrates [82–88].

Mesophilic anaerobic digestion of crude glycerol was studied in work Lopez et al. (2009). The substrate was previously treated in two different ways: (1) acidification with phosphoric acid and centrifugation (so-called acidified glycerol) or (2) acidification followed by distillation (so-called distilled glycerol) [78]. Either granular sludge from anaerobic reactor treating brewery wastewater or nongranular sludge from anaerobic reactor treating urban wastewater was used for inoculation of batch laboratory-scale reactors, having the working volume of one liter. The variations in the methane production were studied, considering the different ways of substrate pretreatment and different types of sludge. The use of the combination of granular sludge with acidified glycerol was found to be the best option for anaerobic treatment of glycerol [78]. The organic loading rates for each substrate and sludge type were in the range of 0.92–2.0 kg/m$^3$·d (COD). Organic loading rate (ORL) is presented as the weight of organic matter per day applied over a specific volume of reactor. The parameter COD (chemical oxygen demand) represents indirectly the amount of organic compounds in the sample. It is a measure of the oxygen needed to degrade organic matter. A decrease in specific methane production was observed when the ORL was increased further. Considering the biomass production and cell maintenance null, 0.382 m$^3$ of methane are theoretically produced per kilogram of removed COD. Experimentally, the effectiveness of the process in each case was: 76% using granular sludge-acidified glycerol, 75% using nongranular sludge-acidified glycerol and 93% with granular sludge-distilled glycerol (0.292; 0.288 and 0.356 m$^3$/kg COD removed, resp.). Besides the methane production coefficient, the removed COD percentage is also important in order to determine biodegradability. This was found to be around 100% using granular sludge-acidified glycerol, 75% with nongranular sludge-acidified glycerol and 85% using granular sludge-distilled glycerol.

Crude glycerol was processed in anaerobic laboratory-mixed reactor under mesophilic conditions for several moths by Bodík et al. [79]. The anaerobic reactor achieved stable operation at the volume loading of 4 kg/m$^3$d with biogas production ca0.980 m$^3$/L of dosed g-phase. The maximal reached volumetric loading was 8–10 kg/m$^3$d, but the loading was considered to be very sensitive and unstable, because it caused decrease of the specific methane production and increase of concentrations of volatile fatty acids (VFA) and dissolved COD. Very effective transformation of g-phase into biogas was measured (more than 95%) which gives very good assumptions for posttreatment of sludge water. The concentration of dissolved inorganic substances increased during the monitored period very slowly but continuously from 1.3 g/L up to 15 g/L. Higher concentrations of dissolved salts could cause inhibition of anaerobic degradation, however no significant influence was observed during this experiment.

In the work Hutňan et al. (2009) results of crude glycerol treatment in the laboratory-mixed reactor (with effective volume of 4 liters) and in the laboratory UASB (upflow anaerobic sludge blanket) reactors (volume of 3.7 liters) are described [80]. From this work resulted that the operation of mesophilic anaerobic degradation of crude glycerol as the only organic substrate is feasible, however the process operation is very sensible to organic overloading of reactor. The laboratory-mixed reactor achieved stable operation at ORL of 4 kg/m$^3$d (COD). The specific production of biogas achieved ca0.980 m$^3$/L of glycerol added. The laboratory UASB reactor with granulated biomass achieved stable operation at ORL of 6.5 kg/m$^3$d and the specific biogas production was ca0.840 m$^3$/L of glycerol added. Inoculation of the UASB reactor with suspended biomass showed that this type of sludge is not suitable for this purpose because of sludge flotation during the reactor operation.

Yang et al. (2008) examined biodegradation of glycerol-containing synthetic wastes using a fixed-bed laboratory bioreactor packed with polyurethane under mesophilic and thermophilic anaerobic conditions [81]. Better performance was obtained from the reactor under the thermophilic conditions. When increasing the ORL from 0.25 to 1 kg/m$^3$d, the COD removal efficiency was decreasing under mesophilic conditions, however under thermophilic conditions higher COD removal was achieved corresponding to higher loading rate. After 516 days of reactor operation, the bed materials under the thermophilic reactors were removed to measure the quantity of attached biomass and for microscopic observation. The polyurethane immobilization carrier retained more biomass than did the liquid phase of reactor. About 95% of the microbes were maintained on the fixed-bed. The immobilized microorganisms present in the thermophilic reactor were primarily Methanobacterium sp., Methanosarcina sp., Bacillus sp., Clostridium sp., Desulfotomaculum sp. andRuminococcus.

Feasibility of utilization of crude glycerol as a cosubstrate has been proven for example in the work of Fountoulakis (2009). The effects of g-phase on the performance of anaerobic reactor treating different types of organic waste (organic fraction of municipal solid waste, mixture of olive mill wastewater and slaughterhouse wastewater) were examined, in order to enhance methane production and increase the yield of hydrogen [82]. Digestion was carried out in a single-stage reactor with a working volume of 3 liters, inoculated with anaerobic sludge from municipal sewage treatment plant and the share of crude glycerol made 1% (v/v) of the dose. The supplementation of the feed with crude glycerol had a significant positive effect and the methane production rate in both cases increased close to the theoretical values given total biodegradation of glycerol. Addition of g-phase to a reactor treating the organic fraction of municipal solid waste resulted in the increase of methane production to 2.094 L/d, compared to 1.400 L/d. An enhanced methane production was also observed when a mixture of olive mill wastewater and slaughterhouse wastewater was supplemented with crude glycerol. Specifically, by adding 1% of g-phase to the feed, the methane production rate increased from 0.479 L/d to 1.210 L/d. Stable concentration levels of COD indicated that COD attributed to glycerol in the feed was totally digested in the reactor. The estimated yield of methane generated from the digestion of glycerol was in both cases almost reaching the value of theoretical methane production. The theoretical methane production from digestion of glycerol was estimated at 0.751 L/d (using the Buswell formula and the ideal gas law).

Fountoulakis et al. studied also the feasibility of adding crude glycerol to the anaerobic digesters treating sewage sludge in wastewater treatment plants [83]. Both batch and continuous experiments were carried out at 35°C. It was observed that glycerol addition up to 1%

(v/v) in the feed increased methane production in the reactor above the expected theoretical value, as it was totally digested and furthermore enhanced the growth of active biomass in the system. On the other hand, any further increase of glycerol caused a high imbalance in the anaerobic digestion process. The reactor treating the sewage sludge produced $1.106 \pm 0.036$ L/d of methane before the addition of glycerol and $2.353 \pm 0.094$ L/d after the addition of glycerol (1% in the feed). The extra glycerol-COD added to the feed did not have a negative effect on reactor performance, but seemed to increase the active biomass (volatile solids) concentration in the system. Also, the kinetic experiments have shown that glycerol biodegradation took place significantly faster than propionate (which is an intermediate product) biodegradation, and it was therefore suggested that the glycerol overload in the reactors increased propionate concentration.

Ma et al. studied the improvement of anaerobic treatment of potato processing wastewater in a laboratory UASB reactor by codigestion with crude glycerol [84]. Influence of three types of glycerol was tested: pure glycerol, crude glycerol and high conductivity glycerol. All 3 types of glycerol are generated as a by-product by the production of biodiesel. They are obtained by different processing technologies and thus their characteristics differ. Supplement of pure glycerol of 2 mL/L of potato processing wastewater resulted in increase of the specific biogas production by $0.740$ m$^3$/L of glycerol added. High COD removal efficiencies (around 85%) were obtained. Moreover, a better in-reactor biomass yield (surplus of active biomass in the reactor) was observed for the UASB reactor supplemented with so-called pure glycerol ($0.012$ g VS (volatile solids) per gram of COD removed) compared to the reactor without added glycerol ($0.002$ g VS per gram of COD removed), which suggests a positive effect of glycerol on the sludge blanket growth.

Álvarez et al. carried out a laboratory study, aimed at maximizing methane production by anaerobic codigestion of three agroindustrial wastes: crude glycerol, pig manure and tuna fish waste [85]. Experiments were performed by batch (discontinuous) assays and 500 mL reactors were operated under the temperature of 35°C. Different blends composed by various percentages of these substrates were fed into the reactors. Compositions of these blends were specified using linear programming optimization method to find most suitable ratios of cosubstrates which would achieve highest biodegradation potential or highest methane production rate. The highest biodegradation potential (methane production of 0.321 $m^3$/kg COD) was reached with a mixture composed of 84% pig manure, 5% fish waste and 11% biodiesel waste, while the highest methane production rate (16.4 L/kgd (COD)) was obtained by a mixture containing 88% pig manure, 4% fish waste and 8% biodiesel waste. Mixture composed of 84% pig manure, 5% fish waste and 11% biodiesel waste and mixture of 79% pig manure, 5% fish waste and 16% biodiesel waste have also achieved very high methane production rates (14.4 L/kg·d (COD) and 12.8 L/kg·d (COD) resp.) compared to the control sample using pig manure substrate (8.3 L/kgd (COD)).

Anaerobic codigestion of crude glycerol in the reactors processing maize, maize silage and pig manure as main substrates, is described in work of Amon et al. (2004). Laboratory digesters under mesophilic conditions were processing a basic mixture, which included 31% of maize silage, 15% of corn maize and 54% of pig manure, together with addition of various levels of g-phase (3, 6, 8 and 15%). The methane yield from the basic mixture without glycerine addition reached 0.335 $m^3$/kg VS [86]. Addition of 3% of glycerine increased the methane yield by 20% and achieved 0.411 $m^3$/kg VS. The addition of 6% of glycerine resulted in the highest methane yield of 0.440 $m^3$/kg VS. Addition of more than 6% glycerine to the basic mixture had only a low positive influence on the methane yield. Addition of 15% glycerine even decreased the methane yield to 0.400 $m^3$/kg VS and the duration of fermentation increased. Methane formation at the start of the experiments was delayed. Analysis of the VFA concentrations in the mixture during the experiments resulted in the hypothesis that the inhibition of methane formation was caused by increased concentration of propionic and butyric acids. The large amounts of these acids were built during decomposition of methanol. VFA accumulation reflects a kinetic uncoupling between acid producers and consumers and is typical for stress situations [92]. The main cause of the toxic effects of high VFA concentrations on the anaerobic digestion process is generally considered to be the resulting drop in pH.

Long-term operation of anaerobic digester for cofermentation of maize silage and crude glycerol was studied in work Špalková et al. (2009). Two laboratory models of a volume of 6 liters were fed by maize silage and a mixture of maize silage with crude glycerol and operated under mesophilic conditions [87]. During the operation period, no negative influence of supplementation of the feed with crude glycerol was observed. Biogas production as well as the sludge water quality (pH, concentrations of COD, VFA, ammonia and phosphate) was similar in both reactors. Maximum portion of g-phase added formed 41.5% of total daily COD dose (together with maize silage). Specific biogas production achieved was approximately $0.40\,m^3/kg$ (COD) in the case of both sole maize silage and a mixture of maize silage with g-phase, meaning that both the maize silage and g-phase had similar specific biogas productions per unit quantity of COD.

A positive effect of glycerol as a cofermentation medium is supported by Amon et al. (2006). Biogas productions from pig manure, crude glycerol and a mixture of 94% of manure with 6% of glycerol were compared in the study [88]. A 6% supplementation of glycerol to pig manure and maize silage resulted in a significant increase in methane production from 0.569 to $0.679\,m^3/kg$ (VS). The methane yield of the mixture supplemented with glycerine was higher than the combined methane yields of both substrates if digested separately. Increase in the specific methane production could not be just corresponding to supplemented glycerol, but was also result of the improved anaerobic degradation caused by the effect of codigestion. Co-digestion of various substrates provides in many cases suitable option for anaerobic processing for various technical reasons. One of the main reasons is the stability of pH and sufficient buffer capacity. Lack of nutrients or high concentration of inhibitory agents can also be improved by sensible choice of cosubstrates. A particularly strong reason for codigestion of feedstock is the adjustment of the carbon-to-nitrogen (C/N) ratio. Digestion of hardly degradable substances was found to be faster by the addition of easily degradable substrates. Moreover some previously problematic wastes were found digestible if digested in a mixture of other waste.

The work by Hutňan et al. (2009) showed that crude glycerol is also a suitable cosubstrate in the full scale biogas plant for anaerobic treatment of maize silage [80]. Reactor was operated under mesophilic conditions and the effective volume of a full scale anaerobic reactor was $2450\,m^3$. Evaluated specific production of biogas from crude glycerol was about $0.890\,m^3/kg$ of crude glycerol added. Dose of crude glycerol, which represented only 5.2% of overall dose to biogas plant, produced almost 15% of overall biogas production. A significant influence on positive economical balance of biogas plant using this cosubstrate has been demonstrated in this study. At the electrical power output of cogeneration unit 300 kW is the daily share of electricity produced from crude glycerol 1067 kWh. At the current price of 0.15 per 1 kWh, this represents a daily profit of 156.55   and a saving of almost 15% silage (1865 kg at a price around 60   for 1 ton).

The possible inhibition effects, resulting from the substrate composition, have to be considered by anaerobic treatment of crude glycerol. Metabolism of the anaerobic microorganisms may be negatively affected mainly by the high salinity of g-phase [79, 80]. The relatively high content of sodium or potassium salts (ca 20–100 g/L) originates from the catalyst, used for the biodiesel production. Higher concentrations of sodium in the anaerobic reactor can seriously inhibit the microbial activity [93].

Biological processing of organic materials in the presence of salts have been studied mainly as an alternative possibility of treating wastewater from industrial processes [94–97] (meat canning, pickled vegetables, dairy products, olive and fish processing industries, petroleum, textile and leather industries). The anaerobic digestion of industrial saline effluents, predominantly from seafood processing, at salt concentrations ranging from 10 to 71 g/L has been studied recently,using different processes, such as an anaerobic filter, UASB reactor and an anaerobic contact system [94, 95, 98]. The COD removal efficiencies obtained, generally remained between 70% and 90%, with OLR ranging from 1 to 15 kg/m³d (COD).

The concentration of sodium exceeding 10 g/L was for a long time generally considered to strongly inhibit methanogenesis [94]. However, the anaerobic digestion in the high salinity level was proven to be possible for treatment of fish-processing effluent [99, 100], if a suitable strategy for adapting the methanogenic biomass was applied. Furthermore, it was shown, that the toxicity of sodium in sludge depends on several factors, such as the type of methanogenic substrate used, the antagonistic or synergistic effects of other ions, the nature and the progressive adaptation of sludge to high salinity and reactor configuration [93, 101].

These factors may be the cause of different results achieved in different studies. Some of the researchers reported the concentrations from 0.9 g/L to 8 g/L to be slightly inhibitory (reduction in methane production by 10%), using different types of substrates [102]. In other experiments, the concentrations in the range of 5.6–53 g/L, depending on the conditions, have been documented to cause the decrease of methane production to a half [93,101–103].

Adequate adaptation of the sludge appears to be of extreme importance, hence the continuous exposure of methanogenic sludge seams to lead to the tolerance of a higher salinity compared to the sludge exposed to salt shocks. The adaptation includes gradual increase of salt concentrations in the sludge, by low organic loading, providing adequate conditions for internal structural changes in the predominant species of methanogens, to adapt to higher osmolarity [104]. Hence the startup period may take several months [94]. According to Gebauer [97] is the adaptation to high sodium concentrations more likely to happen as a result of selection of tolerant species than by adaptation of every single microorganism.

Methanogenic microorganisms seem to be more affected by sodium toxicity than other populations, such as propionate utilizes [101]. The most sensitive appeared to be the nitrogen removing microorganisms [98]. Provided the biomass is acclimated, high salinity is reportedly not an obstacle to its growth and it has no negative influence on sedimentation properties of the sludge or the granulated sludge viscosity. The lack of macronutrients (nitrogen, phosphorus, sulfur) in the medium was found to have a more pronounced negative effect on biomass under the saline conditions. On the other hand, the absence of micronutrients did not further reduce biomass activity under salinity [101, 105].

Another important concern about the anaerobic digestion of crude glycerol should be the concentration of nitrogen-rich substances. Ammonium concentration up to 200 mg/L in the anaerobic reactor is considered to be beneficial [93], since nitrogen is an essential nutrient for microorganisms. Considering the low concentration of nitrogen in the crude glycerol, it may be necessary to supply the nitrogen-rich substances into the reactor. Urea or $NH_4Cl$ are most frequently used as external source of ammonium nitrogen.

# OIL CAKES

Oil cakes or oil meals are solid residues obtained after oil extraction from the seeds. Their composition widely varies depending on the quality of seeds or nuts, growing conditions and extraction methods. Oil cakes can be either edible or nonedible. Edible cakes have a high protein content ranging from 15 to 50%. The chemical compositions of oil cakes originating from different types of plants are listed in Table 3 [106].

**Table 3**: Composition of oil cakes

| Oil cake | Dry matter % | Crude protein % | Crude fibre % | Ash % | Calcium % | Phosphorus % |
|---|---|---|---|---|---|---|
| Canola oil cake | 90 | 33.9 | 9.7 | 6.2 | 0.79 | 1.06 |
| Coconut oil cake | 88.8 | 25.2 | 10.8 | 6.0 | 0.08 | 0.67 |
| Cottonseed cake | 94.3 | 40.3 | 15.7 | 6.8 | 0.31 | 0.11 |
| Groundnut oil cake | 92.6 | 49.5 | 5.3 | 4.5 | 0.11 | 0.74 |
| Mustard oil cake | 89.8 | 38.5 | 3.5 | 9.9 | 0.05 | 1.11 |
| Olive oil cake | 85.2 | 6.3 | 40.0 | 4.2 | — | — |
| Palm kernel cake | 90.8 | 18.6 | 37 | 4.5 | 0.31 | 0.85 |
| Sesame oil cake | 83.2 | 35.6 | 7.6 | 11.8 | 2.45 | 1.11 |
| Soy bean cake | 84.8 | 47.5 | 5.1 | 6.4 | 0.13 | 0.69 |
| Sunflower oil cake | 91 | 34.1 | 13.2 | 6.6 | 0.30 | 1.30 |

In our geographic area (EU), rapeseed and sunflower are the most frequently used substrates, hence we are focusing on the by-products from their processing.

Depending on the method of oil extraction from the seeds, two basic types of solid by-products are generated. Oil cakes are produced when simple oil pressing system is used. In case that pressing is followed by advanced extraction techniques, residues are usually referred to as oil meals. As can be seen in Table 4, main difference between the oil cakes and oil meals is based on the content of fats. The more effective is the extraction process, the fewer lipids remain in the cakes. About 12% of fats (or even 20% in case of small processing facilities) may remain in the oil cake when simple pressing method is employed. Second pressing, sometimes accompanied by water vapor extraction, can lower the content of fats to approximately 8%. If the extraction using hexane is engaged, oil meal with fats content about 1–3% can be generated.

**Table 4**: Composition of rapeseed, rapeseed cake after extraction of 60, 70, and 75% of oil and rapeseed meal, in percents of total solids [111]

| Feed stock | Rapeseed | Rapeseed cake | | | Rapeseed meal |
|---|---|---|---|---|---|
| Portion of extracted oil | | 60% | 70% | 75% | |
| Crude oil | 45 | 24.7 | 19.7 | 17 | 4.5 |
| Crude protein | 23 | 31.5 | 33.6 | 34.7 | 40 |
| Crude fibre | 7 | 9.6 | 10.2 | 10.6 | 12.3 |
| Ash | 5 | 6.8 | 7.3 | 7.5 | 7.7 |

Oil cakes have been currently in use predominantly for feed applications to poultry, ruminant, fish and swine industry [107–110]. Some of them are considered to be suitable organic nitrogenous fertilizers. Several cakes have been utilized for production of proteins, enzymes, antibiotics, mushrooms, ethanol [107–112]. Biotechnological applications of oil cakes also include production of vitamins and antioxidants [111, 113].

Current prices of oil cakes are relatively high (rapeseed cake and meal are in Europe worth approximately 166 and 161 Euros per ton, resp.), compared to other agroindustrial by-products and wastes, which could be also used as substrates for biogas production. However experts are warning against their expected drop due to possible overproduction [114–116]. Moreover, with increasing emphasis on cost reduction of industrial processes and value addition to agroindustrial residues, alternative utilization for oil cakes has been required.

Utilization of oil cakes as an energy source is under examination for now. Some of the oil cakes have been studied as possible feedstocks for biogas production, combustion or pyrolysis [117]. Considering the high content of fats, oil cakes have a high energetic value. They could be suitable substrates for combustion, however because of the large quantity of ash and high emissions of nitrogen oxides, advanced purification technology is required.

Oil cakes and meals contain a high portion of digestible substances, which makes them suitable substrates for the production of biogas. Nutritional content should not be significantly affected by the anaerobic degradation (nutrients such as nitrogen and phosphorus stay in the digestate after degradation) and the digestate should be a convenient agricultural fertilizer. In addition, the plant nutrients contained in cakes are more easily available after the biodigestion.

# ANAEROBIC DIGESTION OF RAPESEED OIL CAKE

Rapeseed cake and rapeseed meal are degradable organic substances. They are suitable for anaerobic digestion, however supplementation of other organic substrates might be required to achieve better process performance and particular problems of digestion should be more closely studied.

Rapeseed cake is a protein-rich substrate, hence the decomposition and conversion to biogas takes longer time than decomposition of substrates rich in carbohydrates. In case of protein degradation, hydrolysis is the limiting step, specifically the cracking of proteins into amino acids and polypeptides by extracellular enzymes. The hydrolysis of carbohydrates takes place within a few hours, while the hydrolysis of proteins within few days. Rate and readiness of degradation of different types of carbohydrates can quite vary. Fats are often decomposed completely. Hemicellulose and lignin, forming the shells of rapeseed, could be quite difficult to decompose in the process of biogas production.

Accumulation of free fatty acids can cause a problem by digestion of materials with higher content of oil, considering that the fats decomposition step is faster than the methanogenesis. Generally, hydrolytic and acidogenic microorganisms are growing about ten times faster than methanogens. Co-digestion of rapeseed cake or meal with other feedstocks, such as manure, provides an alternative solution. Improvement of the biogas production is expected, based on the high oil content.

Rapeseed cake and rapeseed meal are nitrogen-rich media, they content about 35–40% of nitrogen substances [106]. These substances are predominantly proteins, containing amino acids. Expressed in the terms of carbon-to-nitrogen ratio, this makes about 5–8 in case of rapeseed cake. Compared to other materials, the C/N ratio of lignocellulosic materials is in the range of 60–400, grass and silages have C/N about 20–40, swine manure about 12–15 and sunflower cake about 12-13 [118]. The C/N ratio of substrate is an important parameter to be considered by anaerobic degradation, since high content of nitrogen may cause too high content of ammonium nitrate in the biogas reactor. Ammonium nitrogen levels of about 4 g/L of wet sludge bring the risk of process inhibition. If the ammonium content is too high, it is necessary to dilute the substrate with water or nitrogen-poor material.

Phytotoxic effects of rapeseed cake, caused by the content of glucosinolates, must also be considered. They play an important role in the process of digestion, since in higher concentrations they may have harmful effect on methanogenic. The risk of inhibition is getting less serious with decreasing of the glucosinolates level in the rapeseed meal and cake.

There is not much information about experimental anaerobic processing of rapeseed cake or meal in the available literature.

Bernesson et al. estimated the potential biogas production from rapeseed, rapeseed meal and rapeseed cake after 60–75% extraction of oil [111]. Table 5 indicates, that with the increased amount of oil extracted in the process, the possible biogas production from rapeseed cake is decreasing.

**Table 5**: Methane production potential from rapeseed, rapeseed cake and meal in m³/kg VS [111]

| Feed stock | Rapeseed | Rapeseed cake | | | Rapeseed meal |
|---|---|---|---|---|---|
| Portion of extracted oil | | 60% | 70% | 75% | |
| Saccharides | 0.11 | 0.16 | 0.17 | 0.17 | 0.20 |
| Lipids | 0.43 | 0.24 | 0.19 | 0.16 | 0.04 |
| Proteins | 0.12 | 0.16 | 0.17 | 0.18 | 0.20 |
| Together | 0.66 | 0.55 | 0.53 | 0.51 | 0.45 |
| Together (kg) | 0.47 | 0.39 | 0.37 | 0.36 | 0.32 |
| Calorific value (MJ) | 23.4 | 19.5 | 18.6 | 18.1 | 15.8 |

Antonopoulou et al. carried out batch mesophilic biochemical methane potential tests using rapeseed and sunflower residues as a substrate [119]. The experiments indicated that the biological methane potential of rapeseed and sunflower meal were 0.450 m³/kg and 0.481 m³/kg, respectively. Compared to commonly used substrate maize silage, the potential of these oil meals are about 40% higher, so it suggests interesting substrates for the production of biogas. Various pretreatment methods, such as thermal, chemical (through alkali or acid addition) or combination of the above methods were also tested in the effort to enhance the methane productivity and yield. Thermal pretreatment method was conducted at 121°C for 60 minutes in a pressure cooker. Acid or alkali pretreatment of the feedstocks was conducted by the addition of 2% w/v $H_2SO_4$ or NaOH, respectively, for 60 minutes at a temperature of 25°C or at 121°C for 60 min in a pressure cooker (thermal acid or thermal alkali pretreatment). The experiments showed that the pretreatment methods tested did not enhance the methane potential of the rapeseed and sunflower residues. This could be attributed to the inhibitory compounds which were possibly released during the pretreatment.

# ANAEROBIC DIGESTION OF SUNFLOWER OIL CAKE

Sunflower oil cakes and meals are also feasible feedstocks for anaerobic digestion. Raposo et al. examined their anaerobic degradability, biochemical methanogenic potential and the influence of substrate to inoculum ratio in batch laboratory-scale digesters [120, 121]. High stability of the anaerobic digestion process of sunflower oil cake under mesophilic conditions was demonstrated.

The experimental study, with the duration of 7 days, was carried out in a multibatch reactor system [120, 121], which consisted of continuously stirred flasks with an effective volume of 250 mL. The six different inoculum to substrate ratios were tested: 3.0, 2.0, 1.5, 1.0, 0.8 and 0.5. The ultimate methane yield decreased considerably with the inoculum to substrate ratio. The yield of methane was in the range from 0.227 $m^3$/kg for the ratio of 3.0 to 0.107 $m^3$/kg (VS) for the ratio of 0.5. Biodegradability copied this trend, from 86% to 41% was achieved. Higher contribution of substrate may cause lower methane yield due to higher energy consumption in the hydrolytic-acidogenic stage. However, the net VS removed only varied from 42% to 36%, when the ratio decreased from 3.0 to 0.5, which demonstrated the adequate operation of the hydrolytic-acidogenic stage.

The increase in CODs concentrations presented 780–6100 mg/L for the inoculum to substrate ratios of 3.0–0.5, respectively. In case of inoculum-substrate ratio of 3.0 or 2.0, the final values of total VFA were proportional to the amount of substrate added, and no accumulation occurred. However, when the ratio was lower than 2.0, an imbalance of the process was observed, when the VFAs increased to 2050 or 5500 mg/L (for ratios 0.8 and 0.5, resp.). The dissolved CODs also increased, reaching the levels of 3380–12100 mg/L after seven days for the inoculum-substrate ratios of 3.0–0.5, respectively. The trend in the increase of COD with digestion time observed was due mainly to the accumulation of VFA, which reflects a kinetic uncoupling between acid formers and consumers and is typical for a stress situation. This means that the hydrolytic-acidogenic stage was carried out satisfactorily and the imbalance of the process was due to the stress of methanogenic microorganisms.

The net production of total ammonia nitrogen increased with the load added, as a consequence of degradation of proteinsfrom the sunflower oil cake, achieving a maximum value of 1085 mg/L at the ratio of 0.5 (198 mg/L at the ratio 3.0). However, the specific total ammonia nitrogen reached in all ratios similar production of about 40 mg/g VS added.

Identification of the individual VFA may also provide valuable information on the metabolic pathways involved in the process. The high influence of inoculum-substrate ratio on the composition and concentration of the different VFA was shown. By the inoculum to substrate ratio of 3 and 2, the predominant VFA were valeric and butyric acids, but the residual compound was the latter. The absence of acetic and propionic acids indicated, that the methanogenic stage was not disturbed and the formation of methane from these intermediates was quick.

When the ratios of 1.5, 1.0 and 0.8 were applied, the predominant VFA during the first few days were acetic and propionic acids, followed by valeric and butyric acids. Although in the end valeric acid dominated. This performance demonstrated that the lower inoculum-substrate ratio causes the greater accumulation of the longer chain VFA.

By the ratio of 0.5 the predominant VFA were acetic and propionic acids during the first few days, followed by a decrease in acetic acid with time, with a significant residual concentration of propionic, valeric and butyric acids. The VFA profile obtained is a consequence of the imbalance in the methanogenic stage.

De La Rubia et al. investigated also influence of the hydraulic retention time (HRT, it is a measure of the length of time that sludge remains in reactor.) and OLR on the performance of the hydrolytic-acidogenic step of a two-stage anaerobic digestion process of sunflower oil cake [122]. The experiments were performed in laboratory-scale completely stirred tank reactors, with a working volume of 2 L, at mesophilic (35°C) temperature. Digesters were operated over a total period of approximately 350 days. Six OLRs (ranging from 4 to 9 kg/$m^3$d(VS)) for four HRTs (8, 10, 12 and 15 days) were tested to check the effect of each operational variable. Hydrolysis yields obtained for all HRTs and OLRs assayed were in the range of 20.5–30.1%.

Variations in HRT did not affect the COD solubilization of this substrate within the HRT range (15–8 days) researched. Variations in OLR affect the organic matter liquefaction slightly, the highest value (30.1%) being achieved for HRT of 10 days and OLR of 6 kg/$m^3$d(VS). The acidification yield increased with OLR up to 6 kg/$m^3$d(VS), the highest value (83.8%) being achieved for HRT of 10 days and an OLR of 6 kg/$m^3$d(VS). However, higher loading provokes a decrease in the acidification yield, probably due to the fact that the acidogenic bacteria could have been affected and inhibited at the highest OLR studied.

# WASHING WATER

Another possible by-product from biodiesel production offers the water, generated by washing of raw biodiesel. Under the conventional process (alkali-catalyzed transesterification) for every 100 L biodiesel produced about 20 L of washing water is discharged(or more in case of prior acid pretreatment) [123].

Washing water (usually referred to as biodiesel wastewater) is a viscous liquid with an opaque white color similar to aqueous soap. It contains significant amounts of methanol, glycerol and soaps. Methyl esters bound with soap, NaOH or KOH from the catalyst, sodium or potassium salts and trace mono, di- and triglycerides bound up with the soap are also contained in the water.

A great variety of systems for biodiesel purification is available commercially and new alternative technologies are also being investigated. The possible options include dry washing. In this case, the impurities from biodiesel (free glycerol, soap, free fatty acids, catalyst,glycerides, etc.) are absorbed to form a solid waste product instead of a liquid. Dry washing replaces water with an ion exchange resin or a magnesium silicate powder. Both these methods are being used in industrial plants [124]. No regeneration is normally applied and the spent material has to be disposed of to landfill or other applications (compost, potential animal feed additive and potential fuel).

Relatively expensive ultrafiltration or reverse osmosis could also be applied for purification of biodiesel. Yet, the washing with water remains the most convenient alternative [125–127].

Besides crude glycerol, oil cakes and biodiesel wastewater, other potential by-products can be generated in the biodiesel industry. These products are specific, depending on the processing technologies in biodiesel plants. For example, some facilities utilize citric acid solutions in order to wash reactors and other equipment, which produces possible additional waste.

Like the raw glycerol, washing water has also high levels of COD, values in the range of 18–800 g/L have been reported [123, 128–130]. High content of degradable organic substances makes it a suitable source of carbon for microbiological processes, however some issues have to be considered. The wastewater is basic (alkaline), due to the significant levels of residual KOH, and contains a high level of oil and grease and has a high solid content. Nutrients for microbial growth (such as nitrogen and phosphorus) are not abundant in washing water, except for the carbon source. Together these components inhibit the growth of most microorganisms making this wastewater difficult to degrade naturally [123]. Focusing on anaerobic degradation, long-chain fatty acids, which are present on a high level in the washing water, have been reported to be inhibitors of the digestion process [93]. To reduce this effect, electrocoagulation has been proposed as a successful pretreatment for oily wastewater with a subsequent anaerobic treatment [129, 130].

With the likely expansion of biodiesel production by plants using the conventional method, comes the inherent need to treat the wastewater. The main component of the wastewater is the residual remaining oil, thus, such wastewater should not be discharged into public drainage because the oil causes plugging of the drainage and decreases biological activity in sewage treatment. Some of the typical commercially available treatments of oily wastewater employ a dissolved air floatation technique or oil and grease trap unit [130]. Currently, several processes have been developed to treat the biodiesel wastewater, such as the use of chemical recovery approach and electrochemical treatment [128, 130], but also the employment of microbiological processes [123, 131–134] and anaerobic digestion [129].

In the work of Jaruwat et al. (2010), the management of raw biodiesel wastewater was carried out at a laboratory scale at ambient temperature by a combined protonation based chemical recovery of biodiesel followed by electrochemical treatment of the residual wastewater [128]. The combined treatment completely removed COD and oil and grease, and reduced BOD (biologic oxygen demand) levels by more than 95%.

In the study, carried out by Chavalparit and Ongwandee (2009), electrocoagulation was adopted to treat the biodiesel wastewater [130]. This study demonstrates that the electrocoagulation process using an aluminum anode and graphite cathode is effective in reducing oil and grease and suspended solids by more than 95% in the washing water.

However, the COD removal is achieved by 55% due to less significant removal of glycerol and methanol. Therefore, the electrocoagulation process is possibly suitable for a primary treatment for biodiesel wastewater and it still requires a further biological treatment process. Authors believe that pretreatment with electrocoagulation followed by a biological treatment process is feasible and competitive compared with evaporation or pure physicochemical treatments. It requires less energy consumption, short process time, no chemical addition and less sludge production.

The biological treatment of washing water was investigated by Suehara et al. (2005). For the microbiological degradation using a 10-L fermentor, oil degradable yeast, Rhodotorula mucilaginosa, was used and the optimum conditions were determined [123]. The pH was adjusted to 6.8 and several nutrients such as a nitrogen source (ammonium sulfate, ammonium chloride or urea), $KH_2PO_4$ and $MgSO_4 \cdot 7H_2O$ were added to the wastewater. To avoid the inhibition of the microbial growth, the raw biodiesel wastewater was diluted with the same volume of water. The optimal initial concentration of yeast extract was 1 g/L and the optimal C/N ratio was between 17 and 68 when using urea as a nitrogen source. Authors suggest this biological treatment system to be useful for small-scale biodiesel production plants, because it is simple and no controllers, except for a temperature, are necessary.

Kato et al. (2005) proposed a continuous-type consortium bioreactor for treatment of washing water [131]. The main component of this reactor was bacteria-fixed ceramic material with high-oil degrading capability. A series of oil decomposition tests was carried out using the consortium system, in which the most important bacteria types were Acinetobacter, Bacillus and Pseudomonas. The optimal conditions for operation were confirmed by batch tests: air agitation, pH of around 6 and water temperature of 30°C. This reactor operated almost maintenance-free for one year. The field test results for washing water showed that oil and grease concentrations decreased from an initial 120 g/L to a treated range of 10–30 mg/L.

Papanikolaou et al. investigated valorization of soaps from washing water for the production of microbial lipids of specific structure [132–134]. Several oleaginous yeasts and molds are able to accumulate in abundance storage lipid, and at the same time modify the composition of the fat utilized as the carbon source. In the case of various crude fats or fatty wastewaters of low value, this may be an industrially and financially interesting approach. Potential production of cocoa-butter substitute by Yarrowia lipolytica was studied and the cell growth and lipid accumulation of Y. lipolytica was investigated [132].

The anaerobic codigestion of glycerol and wastewater derived from biodiesel manufacturing was studied in batch laboratory-scale reactors of 1 L volume, inoculated by granular biomass, at mesophilic (35°C) temperature [129]. The main purpose of this study was to evaluate the performance, stability, biodegradability, methane yield coefficient, kinetics of methane production, inoculum-substrate ratio and OLR of the anaerobic codigestion of these by-products derived from biodiesel manufacturing.

Prior to the biological treatment, glycerol was acidified with $H_3PO_4$ in order to recover the alkaline catalyst employed in the transesterification reaction (KOH) as agricultural fertilizer. Wastewater was subjected to an electrocoagulation process in order to reduce its oil content. The pretreated washing water was mixed with glycerol at a proportion of 85–15 (COD), until obtaining a final soluble COD of 300 g/L. After mixing, the anaerobic revalorization of the wastewater was studied employing inoculum-substrate ratios ranging from 5.02 to 1.48 kilogram of VSS (volatile suspended solids) per kilogram of COD and OLR of 0.27–0.36 kg/kgd (COD/VSS). Biodegradability was found to be around 100%, while the methane yield coefficient was 0.310 m³/ kg COD removed. The results showed that anaerobic codigestion reduces the clean water and nutrient requirement, with the consequent economical and environmental benefit.

# CONCLUSIONS

The process of biodiesel production is predominantly carried out by catalyzed transesterification. Besides desired methylesters, this reaction provides also a few other products, including crude glycerol, oil-pressed cakes and washing water. Although their composition widely varies depending on the parameters and substrates used for biodiesel production, all these by-products provide valuable feedstocks for biogas generation. The possibility and performance of anaerobic digestion of these materials have been studied to various extents.

The results can be summarized in few points: (i)Crude glycerol from biodiesel production was proven to be a suitable substrate for anaerobic degradation. A couple of studies have demonstrated the possibility of biogas production, using g-phase as a single substrate.(ii)G-phase has also shown a great potential as a cosubstrate by anaerobic treatment of different types of organic waste: organic fraction of municipal solid waste, mixture of olive mill wastewater and slaughterhouse wastewater and potato processing wastewater. Positive effect of crude glycerol on the enhancement of anaerobic processes was observed by treatment of corn maize, maize silage, and swine manure. (iii)Oil cakes and oil meals can also be used as feasible and economically interesting substrates for biogas production. The possibility of methane production from rapeseed and sunflower oil cakes, which deserve the most interest in our area, has been suggested lately. (iv)Tests of anaerobic degradability, biochemical methanogenic potential and influence of substrate to inoculum ratio demonstrated high stability of the anaerobic digestion of sunflower oil cake under mesophilic conditions. (v)The potential biogas production from rapeseed meal and rapeseed cake was estimated. It was shown, that with the increased amount of oil gained in the extraction process, the possible biogas production from rapeseed cake decreases. (vi)No significant effect of the pretreatment (thermal and chemical), of neither sunflower nor rapeseed residues, on the enhancement of methane yield has been observed so far. (vii)Washing water from biodiesel purification is also a promising material for anaerobic degradation, considering the high content of readily degradable organic substances. However, the possibility of biogas generation has not been sufficiently studied. (viii)The specific inhibition effects, resulting from the substrates composition, have to be considered by anaerobic treatment of biodiesel by-products. In case of anaerobic digestion of crude glycerol, high salinity of the substrates may negatively affect the methanogenic microorganisms. The concentration of ammonium should also be monitored. Since nitrogen is an essential nutrient for microorganisms, the low concentration in the crude glycerol and washing water has to be compensated by ammonium supplement. On the other hand, rapeseed cake contains a high portion of nitrogen-rich substances, which may cause inhibition of digestion due to ammonium accumulation in the reactor.

Utilization of the by-products as a potential source of energy, rather than treat them as a waste, seems to be a convenient way of lowering the costs of biodiesel and making it more competitive.

# ACKNOWLEDGMENTS

This work was supported by the Slovak Grant Agency for Science VEGA (Grant 1/0145/08).

# REFERENCES

1.  J. Janaun and N. Ellis, "Perspectives on biodiesel as a sustainable fuel," Renewable and Sustainable Energy Reviews, vol. 14, no. 4, pp. 1312–1320, 2010

2.  European Biodiesel Board: 2009-2010, EU Biodiesel Industry Restrained Growth in Challenging Times, 2010.

3.  A. Demirbas, "Comparison of transesterification methods for production of biodiesel from vegetable oils and fats," Energy Conversion and Management, vol. 49, no. 1, pp. 125–130, 2008

4.  J. Van Gerpen, "Biodiesel processing and production," Fuel Processing Technology, vol. 86, no. 10, pp. 1097–1107, 2005. View at Publisher

5.  D. Y. C. Leung, X. Wu, and M. K. H. Leung, "A review on biodiesel production using catalyzed transesterification," Applied Energy, vol. 87, no. 4, pp. 1083–1095, 2010

6.  M. Fangrui and A. Milford, "Biodiesel production: a review," Bioresource Technology, vol. 70, no. 1, pp. 1–15, 1999. View at Publisher

7.  M. Balat and H. Balat, "Progress in biodiesel processing," Applied Energy, vol. 87, no. 6, pp. 1815–1835, 2010

8.  F. I. Gomez-Castro, V. Rico-Ramirez, J. G. Segovia-Hernandez, and S. Hernandez, "Feasibility study of a thermally coupled reactive distillation process for biodiesel production," Chemical Engineering and Processing, vol. 49, no. 3, pp. 262–269, 2010.

9.  J. S. Lee and S. Saka, "Biodiesel production by heterogeneous catalysts and supercritical technologies,"Bioresource Technology, vol. 101, no. 19, pp. 7191–7200, 2010. View at Publisher

10. J. Ye, S. Tu, and Y. Sha, "Investigation to biodiesel production by the two-step homogeneous base-catalyzed transesterification," Bioresource Technology, vol. 101, no. 19, pp. 7368–7374, 2010. View at Publisher

11. F. E. Kiss, M. Jovanovi , and G. C. Boškovi , "Economic and ecological aspects of biodiesel production over homogeneous and heterogeneous catalysts," Fuel Processing Technology, vol. 91, no. 10, pp. 1316–1320, 2010. View at Publisher

12. Z. Helwani, M. R. Othman, N. Aziz, W. J. N. Fernando, and J. Kim, "Technologies for production of biodiesel focusing on green catalytic techniques: a review," Fuel Processing Technology, vol. 90, no. 12, pp. 1502–1514, 2009

13. E. Li and V. Rudolph, "Transesterification of vegetable oil to biodiesel over MgO-functionalized mesoporous catalysts," Energy and Fuels, vol. 22, no. 1, pp. 145–149, 2008.

14. G. Arzamendi, I. Campo, E. Arguiñarena, M. Sánchez, M. Montes, and L. M. Gandía, "Synthesis of biodiesel with heterogeneous NaOH/alumina catalysts: comparison with homogeneous NaOH,"Chemical Engineering Journal, vol. 134, no. 1–3, pp. 123–130, 2007.

15. W. M. Antunes, C. D. O. Veloso, and C. A. Henriques, "Transesterification of soybean oil with methanol catalyzed by basic solids," Catalysis Today, vol. 133–135, no. 1–4, pp. 548–554, 2008.

16. X. Liu, H. He, Y. Wang, S. Zhu, and X. Piao, "Transesterification of soybean oil to biodiesel using CaO as a solid base catalyst," Fuel, vol. 87, no. 2, pp. 216–221, 2008.

17. G. Ondrey, "Biodiesel production using a heterogeneous catalyst," Chemical Engineering, vol. 111, no. 11, p. 13, 2004.

18. A. Bajaj, P. Lohan, P. N. Jha, and R. Mehrotra, "Biodiesel production through lipase catalyzed transesterification: an overview," Journal of Molecular Catalysis B, vol. 62, no. 1, pp. 9–14, 2010.

19.   T. Tan, J. Lu, K. Nie, L. Deng, and F. Wang, "Biodiesel production with immobilized lipase: a review,"Biotechnology Advances, vol. 28, no. 5, pp. 628–634, 2010

20.   S. Al-Zuhair, A. Almenhali, I. Hamad, M. Alshehhi, N. Alsuwaidi, and S. Mohamed, "Enzymatic production of biodiesel fromused/ waste vegetable oils: design of a pilot plant," Renewable Energy. In press.

21.   P. D. Patil, V. G. Gude, A. Mannarswamy et al., "Optimization of direct conversion of wet algae to biodiesel under supercritical methanol conditions," Bioresource Technology, vol. 102, no. 1, pp. 118–122, 2011.

22.   S. Saka, Y. Isayama, Z. Ilham, and X. Jiayu, "New process for catalyst-free biodiesel production using subcritical acetic acid and supercritical methanol," Fuel, vol. 89, no. 7, pp. 1442–1446, 2010.

23.   R. Sawangkeaw, K. Bunyakiat, and S. Ngamprasertsith, "A review of laboratory-scale research on lipid conversion to biodiesel with supercritical methanol (2001–2009)," Journal of Supercritical Fluids, vol. 55, no. 1, pp. 1–13, 2010.

24.   A. Demirbas, "Biodiesel from waste cooking oil via base-catalytic and supercritical methanol transesterification," Energy Conversion and Management, vol. 50, no. 4, pp. 923–927, 2009

25.   J. C. Thompson and B. B. He, "Characterization of crude glycerol from biodiesel production from multiple feedstocks," Applied Engineering in Agriculture, vol. 22, no. 2, pp. 261–265, 2006

26.   T. Kocsisová and J. Cvengoš, "G-phase from methyl ester production-splitting and refining," Petroleum and Coal, vol. 48, pp. 1–5, 2006.

27.   L. Bournay, D. Casanave, B. Delfort, G. Hillion, and J. A. Chodorge, "New heterogeneous process for biodiesel production: a way to improve the quality and the value of the crude glycerin produced by biodiesel plants," Catalysis Today, vol. 106, no. 1–4, pp. 190–192, 2005.

28.   N. Pachauri and B. He, "Value added utilization of crude glycerol from biodiesel production: a survey of current research activities," in Proceedings of the ASABE Annual International Meeting, 2006.

29.  A. Singhabhandhu and T. Tezuka, "A perspective on incorporation of glycerin purification process in biodiesel plants using waste cooking oil as feedstock," Energy, vol. 35, no. 6, pp. 2493–2504, 2010.

30.  J. A. Weber, The economic feasibility of community-based biodiesel plants, M.S. thesis, University of Missouri, Columbia, 1993.

31.  N. Luo, X. Fu, F. Cao, T. Xiao, and P. P. Edwards, "Glycerol aqueous phase reforming for hydrogen generation over Pt catalyst—effect of catalyst composition and reaction conditions," Fuel, vol. 87, no. 17-18, pp. 3483–3489, 2008.

32.  D. Pyle, Use of biodiesel-derived crude glycerol for the production of omega-3 polyunsaturated fatty acids by the microalga Schizochytriumlimacinum, Ph.D. thesis, Biological Systems Engineering, 2008.

33.  Z. Chi, D. Pyle, Z. Wen, C. Frear, and S. Chen, "A laboratory study of producing docosahexaenoic acid from biodiesel-waste glycerol by microalgal fermentation," Process Biochemistry, vol. 42, no. 11, pp. 1537–1545, 2007.

34.  G. P. da Silva, M. Mack, and J. Contiero, "Glycerol: a promising and abundant carbon source for industrial microbiology," Biotechnology Advances, vol. 27, no. 1, pp. 30–39, 2009.

35.  Y. Feng, Q. Yang, X. Wang, Y. Liu, H. Lee, and N. Ren, "Treatment of biodiesel production wastes with simultaneous electricity generation using a single-chamber microbial fuel cell," Bioresource Technology, vol. 102, no. 1, pp. 411–415, 2010.

36.  T. Valliyappan, N. N. Bakhshi, and A. K. Dalai, "Pyrolysis of glycerol for the production of hydrogen or syn gas," Bioresource Technology, vol. 99, no. 10, pp. 4476–4483, 2008. View at Publisher

37.  F. Freitas, V. D. Alves, J. Pais et al., "Production of a new exopolysaccharide (EPS) by Pseudomonas oleovorans NRRL B-14682 grown on glycerol," Process Biochemistry, vol. 45, no. 3, pp. 297–305, 2010.

38.  S. J. Yoon, Y. C. Choi, Y. I. Son, S. H. Lee, and J. G. Lee, "Gasification of biodiesel by-product with air or oxygen to make syngas," Bioresource Technology, vol. 101, no. 4, pp. 1227–1232, 2010.

39. M. Hájek and F. Skopal, "Treatment of glycerol phase formed by biodiesel production," Bioresource Technology, vol. 101, no. 9, pp. 3242–3245, 2010

40. K. Zamzow, T. K. Tsukamoto, and G. C. Miller, "Biodiesel waste fluid as an inexpensive carbon source for bioreactors treating acid mine drainage (California-Nevada, United States)," in Proceedings of the IMWA Symposium: Water in Mining Environments, Cagliari, Italy, 2007.

41. I. Bodík, A. Blš¥áková, S. Sedlá ek, and M. Hut an, "Biodiesel waste as source of organic carbon for municipal WWTP denitrification," Bioresource Technology, vol. 100, no. 8, pp. 2452–2456, 2009. View at Publisher

42. N. A. Krueger, R. C. Anderson, L. O. Tedeschi, T. R. Callaway, T. S. Edrington, and D. J. Nisbet, "Evaluation of feeding glycerol on free-fatty acid production and fermentation kinetics of mixed ruminal microbes in vitro," Bioresource Technology, vol. 101, no. 21, pp. 8469–8472, 2010

43. N. Rahmat, A. Z. Abdullah, and A. R. Mohamed, "Recent progress on innovative and potential technologies for glycerol transformation into fuel additives: a critical review," Renewable and Sustainable Energy Reviews, vol. 14, no. 3, pp. 987–1000, 2010.

44. M. Slinn, K. Kendall, C. Mallon, and J. Andrews, "Steam reforming of biodiesel by-product to make renewable hydrogen," Bioresource Technology, vol. 99, no. 13, pp. 5851–5858, 2008.

45. P. J. Lammers, B. J. Kerr, T. E. Weber et al., "Digestible and metabolizable energy of crude glycerol for growing pigs," Journal of Animal Science, vol. 86, no. 3, pp. 602–608, 2008.

46. A. K. Forrest, R. Sierra, and M. T. Holtzapple, "Effect of biodiesel glycerol type and fermentor configuration on mixed-acid fermentations," Bioresource Technology, vol. 101, no. 23, pp. 9185–9189, 2010

47. S. Papanikolaou, S. Fakas, M. Fick et al., "Biotechnological valorisation of raw glycerol discharged after bio-diesel (fatty acid methyl esters) manufacturing process: production of 1,3-propanediol, citric acid and single cell oil," Biomass and Bioenergy, vol. 32, no. 1, pp. 60–71, 2008.

48. C. E. Nakamura and G. M. Whited, "Metabolic engineering for the microbial production of 1,3-propanediol," Current Opinion in Biotechnology, vol. 14, no. 5, pp. 454–459, 2003.

49. K. K. Cheng, J. A. Zhang, D. H. Liu et al., "Pilot-scale production of 1,3-propanediol using Klebsiella pneumoniae," Process Biochemistry, vol. 42, no. 4, pp. 740–744, 2007.

50. W. D. Deckwer, "Microbial conversion of glycerol to 1,3-propanediol," FEMS Microbiology Reviews, vol. 16, no. 2-3, pp. 143–149, 1995.

51. C. J. Sullivan, D. C. Dehm, E. E. Reich, and M. E. Dillon, "Polyester resins based upon 2-methyl-1,3-propanediol," The Journal of Coatings Technology, vol. 62, pp. 37–45, 1990.

52. C. Forschner, D. E. Gwyn, C. Frisch, J. Wang, P. Jackson, and A. Sendijarevic, Utilization of 1,3-Propanediol in Thermoplastic Polyurethane Elastomers (TPUs), Shell Chemicals Ltd., 1999.

53. U. Witt, R. J. Muller, J. Augusta, H. Widdecke, and W. D. Deckwer, "Synthesis, properties and biodegradability of polyesters based on 1,3-propanediol," Macromolecular Chemistry and Physics, vol. 195, pp. 793–802, 1994.

54. A. P. Zeng and H. Biebl, "Bulk chemicals from biotechnology: the case of 1,3-propanediol production and the new trends," Advances in biochemical engineering/biotechnology, vol. 74, pp. 239–259, 2002. View at Scopus

55. S. Igari, S. Mori, and Y. Takikawa, "Effects of molecular structure of aliphatic diols and polyalkylene glycol as lubricants on the wear of aluminum," Wear, vol. 244, no. 1-2, pp. 180–184, 2000

56. Carlo Erba Reagents: 100th edition of Catalogue Chemicals, Section 2, Green Solvents—Sustainable chemistry, pp. 45–50, 2004.

57. F. Barbirato, E. H. Himmi, T. Conte, and A. Bories, "1,3-propanediol production by fermentation: an interesting way to valorize glycerin from the ester and ethanol industries," Industrial Crops and Products, vol. 7, no. 2-3, pp. 281–289, 1998.

58. B. Solomos, A.-P. Zeng, H. Biebl, H. Schlieker, C. Posten, and W.-D. Deckwer, "Comparison of the energetic efficiencies of hydrogen and oxochemicals formation in Klebsiellapneumoniae

and Clostridium butyricum during anaerobic growth on glycerol," Journal of Biotechnology, vol. 39, pp. 107–117, 1995.

59.   A. Hiremath, M. Kannabiran, and V. Rangaswamy, "1,3-Propanediol production from crude glycerol from jatropha biodiesel process," New Biotechnology, vol. 28, no. 1, pp. 19–23, 2011.

60.   T. Ito, Y. Nakashimada, K. Senba, T. Matsui, and N. Nishio, "Hydrogen and ethanol production from glycerol-containing wastes discharged after biodiesel manufacturing process," Journal of Bioscience and Bioengineering, vol. 100, no. 3, pp. 260–265, 2005.

61.   S. Adhikari, S. D. Fernando, and A. Haryanto, "Hydrogen production from glycerin by steam reforming over nickel catalysts," Renewable Energy, vol. 33, no. 5, pp. 1097–1100, 2008.

62.   M. D. Blankschien, J. M. Clomburg, and R. Gonzalez, "Metabolic engineering of Escherichia coli for the production of succinate from glycerol," Metabolic Engineering, vol. 12, no. 5, pp. 409–419, 2010.

63.   S. Ethier, K. Woisard, D. Vaughan, and Z. Wen, "Continuous culture of the microalgaeSchizochytriumlimacinum on biodiesel-derived crude glycerol for producing docosahexaneoic acid,"Bioresource Technology, vol. 102, pp. 88–93, 2011.

64.   B. S. Fernandes, G. Peixoto, F. R. Albrecht, N. K. Saavedra del Aguila, and M. Zaiat, "Potential to produce biohydrogen from various wastewaters," Energy for Sustainable Development, vol. 14, no. 2, pp. 143–148, 2010.

65.   M. Navrátil, J. Tká , J. Švitel, B. Danielsson, and E. Šturdík, "Monitoring of the bioconversion of glycerol to dihydroxyacetone with immobilized gluconobacter oxydans cell using thermometric flow injection analysis," Process Biochemistry, vol. 36, no. 11, pp. 1045–1052, 2001.

66.   H. Hu and T. K. Wood, "An evolved Escherichia coli strain for producing hydrogen and ethanol from glycerol," Biochemical and Biophysical Research Communications, vol. 391, no. 1, pp. 1033–1038, 2010.

67. A. Kivistö, V. Santala, and M. Karp, "Hydrogen production from glycerol using halophilic fermentative bacteria," Bioresource Technology, vol. 101, no. 22, pp. 8671–8677, 2010.

68. K. O. Yu, S. W. Kim, and S. O. Han, "Engineering of glycerol utilization pathway for ethanol production by Saccharomyces cerevisiae," Bioresource Technology, vol. 101, no. 11, pp. 4157–4161, 2010.

69. Z. Yuan, J. Wang, L. Wang et al., "Biodiesel derived glycerol hydrogenolysis to 1,2-propanediol on Cu/MgO catalysts," Bioresource Technology, vol. 101, no. 18, pp. 7088–7092, 2010

70. Y. Zhu, J. Li, M. Tan et al., "Optimization and scale-up of propionic acid production by propionic acid-tolerant Propionibacterium acidipropionici with glycerol as the carbon source," Bioresource Technology, vol. 101, no. 22, pp. 8902–8906, 2010

71. A. André, P. Diamantopoulou, A. Philippoussis, D. Sarris, M. Komaitis, and S. Papanikolaou, "Biotechnological conversions of bio-diesel derived waste glycerol into added-value compounds by higher fungi: production of biomass, single cell oil and oxalic acid," Industrial Crops and Products, vol. 31, no. 2, pp. 407–416, 2010

72. S. Papanikolaou and G. Aggelis, "Lipid production by Yarrowia lipolytica growing on industrial glycerol in a single-stage continuous culture," Bioresource Technology, vol. 82, no. 1, pp. 43–49, 2002.

73. S. Fakas, S. Papanikolaou, A. Batsos, M. Galiotou-Panayotou, A. Mallouchos, and G. Aggelis, "Evaluating renewable carbon sources as substrates for single cell oil production by Cunninghamella echinulata andMortierella isabellina," Biomass and Bioenergy, vol. 33, no. 4, pp. 573–580, 2009

74. A. Makri, S. Fakas, and G. Aggelis, "Metabolic activities of biotechnological interest in Yarrowia lipolytica grown on glycerol in repeated batch cultures," Bioresource Technology, vol. 101, no. 7, pp. 2351–2358, 2010

75. S. Papanikolaou and G. Aggelis, "Biotechnological valorization of biodiesel derived glycerol waste through production of single cell oil and citric acid by Yarrowia lipolytica," Lipid Technology, vol. 21, no. 4, pp. 83–87, 2009

76. S. Papanikolaou and G. Aggelis, "Lipid production by Yarrowia lipolytica growing on industrial glycerol in a single-stage continuous culture," Bioresource Technology, vol. 82, no. 1, pp. 43–49, 2002

77. S. Papanikolaou, L. Muniglia, I. Chevalot, G. Aggelis, and I. Marc, "Yarrowia lipolytica as a potential producer of citric acid from raw glycerol," Journal of Applied Microbiology, vol. 92, no. 4, pp. 737–744, 2002.

78. J. López, M. Santos, A. Pérez, and A. Martín, "Anaerobic digestion of glycerol derived from biodiesel manufacturing," Bioresource Technology, vol. 100, no. 23, pp. 5609–5615, 2009.

79. I. Bodík, M. Hut an, T. Petheöová, and A. Kalina, "Anaerobic treatment of biodiesel production wastes," in Proceedings of the Book of Lectures on 5th International Symposium on Anaerobic Digestion of Solid Wastes and Energy Crops, Hammamet, Tunisia, 2008.

80. M. Hut an, N. Kolesárová, I. Bodík, V. Špalková, and M. Lazor, "Possibilities of anaerobic treatment of crude glycerol from biodiesel production," in Proceedings of 36th International Conference of Slovak Society of Chemical Engineering, Tatranské Matliare, 2009.

81. Y. Yang, K. Tsukahara, and S. Sawayama, "Biodegradation and methane production from glycerol-containing synthetic wastes with fixed-bed bioreactor under mesophilic and thermophilic anaerobic conditions," Process Biochemistry, vol. 43, no. 4, pp. 362–367, 2008.

82. M. S. Fountoulakis and T. Manios, "Enhanced methane and hydrogen production from municipal solid waste and agro-industrial by-products co-digested with crude glycerol," Bioresource Technology, vol. 100, no. 12, pp. 3043–3047, 2009

83. M. S. Fountoulakis, I. Petousi, and T. Manios, "Co-digestion of sewage sludge with glycerol to boost biogas production," Waste Management, vol. 30, no. 10, pp. 1849–1853, 2010.

84. J. Ma, M. Van Wambeke, M. Carballa, and W. Verstraete, "Improvement of the anaerobic treatment of potato processing wastewater in a UASB reactor by co-digestion with glycerol," Biotechnology Letters, vol. 30, no. 5, pp. 861–867, 2008

85.   J. A. Álvarez, L. Otero, and J. M. Lema, "A methodology for optimising feed composition for anaerobic co-digestion of agro-industrial wastes," Bioresource Technology, vol. 101, no. 4, pp. 1153–1158, 2010.

86.   T. Amon, V. Kryvoruchko, B. Amon, and M. Schreiner, Untersuchungen zur Wirkung von Rohglycerin aus der Biodieselerzeugung als leistungssteigerndes Zusatzmittel zur Biogaserzeugung aus Silomais, Körnermais, Rapspresskuchen und Schweinegülle, Ergebnisbericht Department für Nachhaltige Agrarsysteme Institut für Landtechnik, Universität für Bodenkultur, Wien, Austria, 2004.

87.   V. Špalková, M. Hut an, and N. Kolesárová, "Selected problems of anaerobic treatment of maize silage," in Proceedings of 36th International Conference of Slovak Society of Chemical Engineering, Tatranské Matliare, 2009.

88.   T. Amon, B. Amon, V. Kryvoruchko, V. Bodiroza, E. Pötsch, and W. Zollitsch, "Optimising methane yield from anaerobic digestion of manure: effects of dairy systems and of glycerine supplementation,"International Congress Series, vol. 1293, pp. 217–220, 2006.

89.   A. I. Qatibi, A. Bories, and J. L. Garcia, "Sulfate reduction and anaerobic glycerol degradation by a mixed microbial culture," Current Microbiology, vol. 22, no. 1, pp. 47–52, 1991.

90.   A. I. Qatibi, R. Bennisse, M. Jana, and J.-L. Garcia, "Anaerobic degradation of glycerol by Desulfovibrio fructosovorans and D. carbinolicus and evidence for glycerol-dependent utilization of 1,2- propanediol,"Current Microbiology, vol. 36, no. 5, pp. 283–290, 1998.

91.   C. Gallert and J. Winter, "Bacterial metabolism in wastewater treatment systems," in Environmental Biotechnology. Concepts and Applications, H. J. Jördening and J. Winter, Eds., Wiley-VCH, Weinheim, Germany, 2005.

92.   B. K. Ahring, M. Sandberg, and I. Angelidaki, "Volatile fatty acids as indicators of process imbalance in anaerobic digestors," Applied Microbiology and Biotechnology, vol. 43, no. 3, pp. 559–565, 1995.

93. Y. Chen, J. J. Cheng, and K. S. Creamer, "Inhibition of anaerobic digestion process: a review,"Bioresource Technology, vol. 99, no. 10, pp. 4044–4064, 2008. View at Publisher

94. O. Lefebvre and R. Moletta, "Treatment of organic pollution in industrial saline wastewater: a literature review," Water Research, vol. 40, no. 20, pp. 3671–3682, 2006.

95. G. D. Boardman, J. L. Tisinger, and D. L. Gallagher, "Treatment of clam processing wastewaters by means of upflow anaerobic sludge blanket technology," Water Research, vol. 29, no. 6, pp. 1483–1490, 1995.

96. O. Lefebvre, N. Vasudevan, M. Torrijos, K. Thanasekaran, and R. Moletta, "Anaerobic digestion of tannery soak liquor with an aerobic post-treatment," Water Research, vol. 40, no. 7, pp. 1492–1500, 2006.

97. R. Gebauer, "Mesophilic anaerobic treatment of sludge from saline fish farm effluents with biogas production," Bioresource Technology, vol. 93, no. 2, pp. 155–167, 2004.

98. O. Lefebvre, N. Vasudevan, M. Torrijos, K. Thanasekaran, and R. Moletta, "Halophilic biological treatment of tannery soak liquor in a sequencing batch reactor," Water Research, vol. 39, no. 8, pp. 1471–1480, 2005

99. M. Soto, R. Mendez, and J. M. Lema, "Biodegradability and toxicity in the anaerobic treatment of fish canning wastewaters," Environmental Technology, vol. 12, no. 8, pp. 669–677, 1991.

100. F. Omil, R. Méndez, and J. M. Lema, "Anaerobic treatment of seafood processing waste waters in an industrial anaerobic pilot plant," Water SA, vol. 22, no. 1, pp. 173–181, 1996.

101. G. Feijoo, M. Soto, R. Méndez, and J. M. Lema, "Sodium inhibition in the anaerobic digestion process: antagonism and adaptation phenomena," Enzyme and Microbial Technology, vol. 17, no. 2, pp. 180–188, 1995

102. Y. Liu and D. R. Boone, "Effects of salinity on methanogenic decomposition," Bioresource Technology, vol. 35, no. 3, pp. 271–273, 1991.

103. M. Soto, R. Mendez, and J. M. Lema, "Sodium inhibition and sulphate reduction in the anaerobic treatment of mussel

processing wastewaters," Journal of Chemical Technology and Biotechnology, vol. 58, no. 1, pp. 1–7, 1993.

104. R. D. Sleator and C. Hill, "Bacterial osmoadaptation: the role of osmolytes in bacterial stress and virulence," FEMS Microbiology Reviews, vol. 26, no. 1, pp. 49–71, 2002.

105. I. Vyrides and D. C. Stuckey, "Adaptation of anaerobic biomass to saline conditions: role of compatible solutes and extracellular polysaccharides," Enzyme and Microbial Technology, vol. 44, no. 1, pp. 46–51, 2009

106. S. Ramachandran, S. K. Singh, C. Larroche, C. R. Soccol, and A. Pandey, "Oil cakes and their biotechnological applications—a review," Bioresource Technology, vol. 98, no. 10, pp. 2000–2009, 2007.

107. E. Molina-Alcaide and D. R. Yáñez-Ruiz, "Potential use of olive by-products in ruminant feeding: a review," Animal Feed Science and Technology, vol. 147, no. 1–3, pp. 247–264, 2008

108. K. Franke, U. Meyer, H. Wagner, H. O. Hoppen, and G. Flachowsky, "Effect of various iodine supplementations, rapeseed meal application and two different iodine species on the iodine status and iodine excretion of dairy cows," Livestock Science, vol. 125, no. 2-3, pp. 223–231, 2009

109. T. Mushtaq, M. Sarwar, G. Ahmad et al., "Influence of sunflower meal based diets supplemented with exogenous enzyme and digestible lysine on performance, digestibility and carcass response of broiler chickens," Animal Feed Science and Technology, vol. 149, no. 3-4, pp. 275–286, 2009

110. N. M. Soren and V. R. B. Sastry, "Replacement of soybean meal with processed karanj (Pongamia glabra) cake on the balances of karanjin and nutrients, as well as microbial protein synthesis in growing lamb," Animal Feed Science and Technology, vol. 149, no. 1-2, pp. 16–29, 2009

111. S. Bernesson, "Fields of application for the by-products of extraction and tranesterification of rapeseed oil," Tech. Rep. 1652-3237, Department of Biometry andEngineering, Uppsala, Sweden, 2007.

112. J. M. Cerveró, P. A. Skovgaard, C. Felby, H. R. Sørensen, and H. Jørgensen, "Enzymatic hydrolysis and fermentation of palm

kernel press cake for production of bioethanol," Enzyme and Microbial Technology, vol. 46, no. 3-4, pp. 177–184, 2010

113. T. Vaštag, L. Popovi, S. Popovi, V. Krimer, and D. Peri in, "Production of enzymatic hydrolysates with antioxidant and angiotensin-I converting enzyme inhibitory activity from pumpkin oil cake protein isolate," Food Chemistry, vol. 124, no. 4, pp. 1316–1321, 2011.

114. J. N. Ferrer and E. A. Kaditi, "The EU added value of agricultural expenditure—from market to multifunctionality—gathering criticism and success stories of the CAP," European Parliament, Policy Department on Budgetary Affairs, Brussels, Belgium, 2007.

115. Solar Oil Systems: Rapecake: more than just a by-product, 2007,http://www.solaroilsystems.nl/publicaties/img/1184614337_2007-6-Rapeseed_cake_-_more_than_just_byproduct.pdf.

116. International Fertilizer Industry Association (IFA) Task Force on Bioenergy, "A survey of the anticipated impact of biofuel development on short-, medium- and long-term fertilizer demand," in Proceedings of the 75th IFA Annual Conference, Istanbul, Turkey, 2007.

117. S. Yorgun, S. Sensoz, and O. M. Kockar, "Flash pyrolysis of sunflower oil cake for production of liquid fuels," Journal of Analytical and Applied Pyrolysis, vol. 60, no. 1, pp. 1–12, 2001.

118. F. Straka, P. Jeníek, J. Zábranská, and M. Dohányos, "Anaerobic fermentation of biomass and wastes with respect to sulfur and nitrogen contents in treate materials," in Proceedings of the11th International Waste Management and Landfill Symposium, Cagliari, Italy, 2007.

119. G. Antonopoulou, K. Stamatelatou, and G. Lyberatos, "Exploitation of rapeseed and sunflower residues for methane generation through anaerobic digestion: the effect of pretreatment," in Proceedings of the 2nd International Conference of Industrial Biotechnology, Padua, Italy, 2010.

120. F. Raposo, R. Borja, M. A. Martín, A. Martín, M. A. de la Rubia, and B. Rincón, "Influence of inoculum-substrate ratio on the anaerobic digestion of sunflower oil cake in batch mode: process stability and kinetic evaluation," Chemical Engineering Journal, vol. 149, no. 1–3, pp. 70–77, 2009.

121. F. Raposo, R. Borja, B. Rincon, and A. M. Jimenez, "Assessment of process control parameters in the biochemical methane potential of sunflower oil cake," Biomass and Bioenergy, vol. 32, no. 12, pp. 1235–1244, 2008

122. M. A. De La Rubia, F. Raposo, B. Rincón, and R. Borja, "Evaluation of the hydrolytic-acidogenic step of a two-stage mesophilic anaerobic digestion process of sunflower oil cake," Bioresource Technology, vol. 100, no. 18, pp. 4133–4138, 2009

123. K. I. Suehara, Y. Kawamoto, E. Fujii, J. Kohda, Y. Nakano, and T. Yano, "Biological treatment of wastewater discharged from biodiesel fuel production plant with alkali-catalyzed transesterification,"Journal of Bioscience and Bioengineering, vol. 100, no. 4, pp. 437–442, 2005

124. M. Berrios and R. L. Skelton, "Comparison of purification methods for biodiesel," Chemical Engineering Journal, vol. 144, no. 3, pp. 459–465, 2008.

125. M. Hayyan, F. S. Mjalli, M. A. Hashim, and I. M. AlNashef, "A novel technique for separating glycerine from palm oil-based biodiesel using ionic liquids," Fuel Processing Technology, vol. 91, no. 1, pp. 116–120, 2010

126. J. Saleh, A. Y. Tremblay, and M. A. Dubé, "Glycerol removal from biodiesel using membrane separation technology," Fuel, vol. 89, no. 9, pp. 2260–2266, 2010.

127. N. Sdrula, "A study using classical or membrane separation in the biodiesel process," Desalination, vol. 250, no. 3, pp. 1070–1072, 2010

128. P. Jaruwat, S. Kongjao, and M. Hunsom, "Management of biodiesel wastewater by the combined processes of chemical recovery and electrochemical treatment," Energy Conversion and Management, vol. 51, no. 3, pp. 531–537, 2010

129. J. A. Siles, M. A. Martín, A. F. Chica, and A. Martín, "Anaerobic co-digestion of glycerol and wastewater derived from biodiesel manufacturing," Bioresource Technology, vol. 101, no. 16, pp. 6315–6321, 2010.

130. O. Chavalparit and M. Ongwandee, "Optimizing electrocoagulation process for the treatment of biodiesel

wastewater using response surface methodology," Journal of Environmental Sciences, vol. 21, no. 11, pp. 1491–1496, 2009.

131. S. Kato, H. Yoshimura, K. Hirose, M. Amornkitbamrung, M. Sakka, and I. Sugahara, "Application of microbial consortium system to wastewater from biodiesel fuel generator," IEA-Waterqual, Singapore, CD-ROM, 2005.

132. S. Papanikolaou, I. Chevalot, M. Komaitis, G. Aggelis, and I. Marc, "Kinetic profile of the cellular lipid composition in an oleaginous Yarrowia lipolytica capable of producing a cocoa-butter substitute from industrial fats," Antonie van Leeuwenhoek, vol. 80, no. 3-4, pp. 215–224, and 2001.

133. S. Papanikolaou and G. Aggelis, "Selective uptake of fatty acids by the yeast Yarrowia lipolytica,"European Journal of Lipid Science and Technology, vol. 105, no. 11, pp. 651–655, 2003

134. S. Papanikolaou and G. Aggelis, "Modeling lipid accumulation and degradation in Yarrowia lipolyticacultivated on industrial fats," Current Microbiology, vol. 46, no. 6, pp. 398–402, 2003.

# 6

# Sustaining Biodiesel Production via Value-Added Applications of Glycerol

Omotola Babajide[1,2]

[1]Environmental and Nano Sciences Group, Chemistry Department, University of the Western Cape, Private Bag X17, Bellville 7535, South Africa.

[2]School of Sciences, Obafemi Awolowo University, Adeyemi College of Education, PMB 520 Ondo State, Nigeria.

## ABSTRACT

The production of biofuels worldwide has been significant lately due to the shift from obtaining energy from nonrenewable energy (fossil fuels) to renewable sources (biofuels). This energy shift arose as a result of the disturbing crude petroleum price fluctuations, uncertainties about fossil fuel reserves, and greenhouse gas (GHG) concerns. With the production of biofuels increasing considerably and the current global biodiesel production from different feedstock, reaching about 6 billion liters per year, biodiesel production costs have been highly dependent on feedstock prices, ranging from 70 to 25; of total production costs, and in comparison with the conventional diesel fuel, the biodiesel is currently noncompetitive. An efficient production process is, therefore, crucial to lowering biodiesel production costs. The question of sustainability, however, arises, taking into account the African diverse conditions and how vital concerns need to be addressed. The major concern about biodiesel production costs can be reduced by finding value-added applications for its glycerol byproduct. This paper, thus, provides an overview of current research trends that could overcome the major hurdles towards profitable commercialization of biodiesel and also proposes areas of opportunity probable to capitalize the surplus glycerol obtained, for numerous applications.

# INTRODUCTION

Energy plays a vital role in the economic development and social/national security of any nation, as access to secure, sustainable, and affordable energy is a prerequisite for sustainable development [1]. Current patterns of energy supply and energy use are unsustainable because of environmental issues such as global warming strongly associated with increased energy consumption. Energy sufficiency and security is a key factor in development since it provides essential inputs for socioeconomic development that provide vital services which improve the quality of life at regional, national, and subnational levels [2, 3]. Several countries on the African continent continue to face great challenges of energy security and the negative effects of climate change. The strong tie between energy, Millennium Development Goals (MDGs), and widespread poverty makes it important to tackle the challenges and prospects for energy services provision in the continent [4]. The inability of many African countries to provide good and adequate energy services has been a major constraint to their development. The continent remains friable with widespread poverty that is due to several factors [2], and about half of the African population still lives in absolute poverty with approximately 70% depending on traditional biomass as their only source of fuel [5]. Renewable energy technologies (RETs) in general and biofuel specifically offer developing countries some prospect of a self-reliant energy supply at the national and local levels with potential economic, ecological, social, and security benefits [6].

# BIOFUELS DEVELOPMENT IN AFRICA

Africa is an unexplored and unexploited resource for biofuels development. Although the majority of the African countries rely on biomass (mainly wood from forests) as a main energy resource, it is inefficiently and unsustainably used, with detrimental effects on the environment and the well-being of its inhabitants. These detrimental effects are triggered by the main anthropogenic greenhouse gas (GHG), carbon dioxide ($CO_2$), because it is closely associated to the prevalent use of fossil energy. $CO_2$ emissions from the combustion of fossil fuels such as coal, petroleum, and natural gas are the most important sources of anthropogenic GHG emissions all over the industrialized countries. Securing the supply of fossil fuel has besieged and caused environmental destruction of the planet Earth. Energy from renewable sources offers a solution proposed to combat the depletion of the world's nonrenewable energy sources, and this kind of energy is beneficial for the natural environment, but the cost of energy production from renewable sources is very high and exceeds the price of conventional fuels [7–9]. According to a report from the African Development Bank in 2006 [10], about 39% of the total energy consumed in the Subsaharan Africa is imported against a world average of 19% confirming the heavy dependence on imported fuel. In Africa, as reported by Amigun et al. [4], the biodiesel industry is marginal, whereas the production potential is enormous. A number of small-scale plants had sprung up in South Africa, while some large-scale commercial plants in Europe supply and offer consultancy operations which in turn play a significant role in enabling the African corporate partners to achieve the laid down objectives and programs. Since 2005, several African nations have used the private sector investment strategies and now benefit from the establishment of large-scale J. curcas plantations [11]. The most promising cultivation regions are based in Mozambique, Ghana, Malawi, Tanzania, and Zimbabwe [12] with Mozambique widely seen as possessing the largest potential for J. curcas production. The International Energy Agency estimated that Mozambique could produce nearly 3 million barrels of oil a day of liquid biofuels from nonfood crop resources such as J. curcas [13]. In Mozambique, three significant projects were announced during 2006-2007. Canadian-based Energem Resources invested $5.5 million (4 million) on its first small plantation with near-future commitments to cultivate 60,000 hectares. The South Africa-based Duelco renewable energy established Mozambique partnerships around a 60,000-hectare plantation, and ESV-Bio Africa currently manages an 11,000-hectare plantation with plans for 100,000 hectares [14]. Ghana planned a 12,000-hectare project with the South Africa-based BD-1 Group with interest in J. curcas production from Petrobras, and other foreign-based companies. According to a current report, Ghana ranks among the "top 5" list of the developing countries likely to attract biodiesel investment, with this being made possible because Ghana has a unique position of having low debt and perception of corruption [15]. In Tanzania, UK-based Sun Biofuels committed nearly 20 million dollars to J. curcas production and a biodiesel processing plant [14]. There are plans for expansion in other African countries as well, including Malawi, Burkina Faso, and Madagascar giving the increasing support from the private sector stakeholders, lower costs of production, and an ideal growing climate [16]. South Africa's biodiesel market is mainly characterized by several small- and medium-scale producers, while Zimbabwe's six million US-dollar biodiesel plant which was built to process atropha curcas, cotton seed, sunflower, and soya, among other feedstock, has not been fully operational due to political problems currently faced in the country [17]. The use of vegetable oil as a source of fuel for energy production has also been explored in some African countries such as Mali and Uganda. According to reports, South Africa, Angola, and Mozambique have the tendency to become the leading biofuels and carbon credit suppliers to world markets [18]. Currently, Angola has one of the largest nonforest agricultural lands in the world, and it is estimated that the country's biodiesel export potential is estimated at approximately six exajoules of bioenergy per year, the equivalent of 2.7 million barrels of oil per day (bpd) [19]. Angola and Mozambique have also joined forces with the Brazilian companies such as Petrobras (the national oil company), for example, in the development of soybean-based biodiesel allowing for more efficient development of the agriculture biofuels feedstock industry in both countries. Overall, many expect that the biofuels market in Africa will help the continent begin to realize its agricultural potential, bring real investment into irrigation technologies, and also lead to infrastructural growth and development. An attempt to create awareness of a shift to alternative fuel (biodiesel) was shown by a nongovernment, organization, "Journey to Forever", that travelled from Hong Kong to Southern Africa producing their own biodiesel along the way and teaching people from small villages how to make their own biodiesel to use in heaters, tractors, buses, automobiles, and other machines [20]. According to Mwakasonda [21], an energy expert with the United Nations (UN) Environment Program, worldwide investment in bioenergy reached twenty-one billion (USD) in 2008; "African countries are keen on capturing some of this market by transforming their expansive farmlands into the next oil fields." It should be also noted that, while by far the largest quantities of biodiesel fuels are still being produced and marketed within the European Union, the US and the Asian markets, a considerable rate of increase in biodiesel production is projected to occur in Africa with particular interest in South Africa; it is estimated that production might increase in folds by 2015 [22].

# SUSTAINABILITY OF BIOFUELS IN AFRICA

According to Sustainable Development Commission (SDC), the use of biofuels can lead to a reduction in greenhouse gases (GHGs), these reductions, however, require a series of careful measures to make this obtainable and have to be balanced against any environmental and social detriment amongst other factors [10]. There are several reasons for bio-fuels to be considered as relevant technologies by both developing and industrialized countries [2]; these reasons include the: source of foreign exchange saving activity especially for oil-deprived countries (development and use of locally produced renewable fuel and reduction of demand for imported petroleum), boosting of local agriculture productions, and additional markets and revenue to farmers leading consequently to the increase of rural folk's purchasing powers and quality of life, beneficial environmental impact through the usage of organic municipal solid waste materials to generating a higher-value endproduct, reduced level of carbon dioxide emitted by motor engines, and then preservation of the quality of the atmosphere. Ideally, biofuel alternatives should reduce the dependence on oil and contribute as much as possible to meet the GHG emissions target. However, it is also widely accepted that joint efforts from politicians, regulators, scientists, and consumers will be needed to support an independent oil/GHG-controlled scenario in the future. In South Africa, transport use contributes about 16% of the total greenhouse gas emissions, the greatest in Africa, and the transport sector accounts for about 26% of global $CO_2$ emissions [23] of which roughly two thirds originate from the wealthiest 10% of the countries [24], and of this, diesel fuel contributes about 17% or 11,705,000 tonnes of $CO_2$ equivalent; 1,622,000 additional tonnes are released from diesel fuel used for electricity generation [25]. On top of greenhouse gas emissions is the vexing question of how little—or much—is left in order to reduce $CO_2$ emissions from road transport. A significant modal shift onto a more environmentally benign source of energy is required [26] in which oils of vegetable and animal origin, unlike fossil fuels, have the potential to provide not only on a sustainable basis but could also be greenhouse gas neutral, or at the very least, emit substantially less greenhouse gases per unit energy. The need for a secure energy supply for transportation makes it essential to explore biofuels as alternatives to mineral oil-based fuels addressing and evaluating socioeconomic and environmental consequences originated in their implementation. Due to the environmental merits which also include availability from common biomass sources, representation of carbon dioxide cycle in combustion, possession of a considerable environmentally friendly potential, biodegradability, and contributions to sustainability, the share of biofuel in the automotive fuel market has been predicted to grow fast in the next decade according to reports by Monfort [27]. An exponential increase in the consumption of these biofuels has taken place in the last few years in the European Union (EU) according to reports by the IEA, 2007, and in order for biofuels to be viable alternatives to conventional fuels, it is recommended to provide net energy gain and to be environmentally beneficial, economically competitive and available in large quantities without reducing food supplies [28]. With a large landmass for farming, the Subsaharan Africa is increasingly being viewed as a region with a fairly high potential for biofuels production; bearing mainly energy security in mind, a number of countries have begun their own national biodiesel programs (Table 1) in which these countries still lag behind in biofuel potential. The growing interest in biofuels in many African countries can be explained by many factors. High crude oil prices, fluctuations in prices often due to geopolitical uncertainties, local and global environmental impacts of fossil fuels such as climate change, economic development, and employment are some of the drivers that are increasing the widespread interest in biofuels.

**Table 1:** Biofuels production potential of some African countries in megaliters [4]

| Country | Feedstock | Biofuels potential (ML) |
|---|---|---|
| Kenya | Molasses | 413 |
| Ethiopia | Molasses | 80 |
| Nigeria | Sugarcane/Jatropha | 70 |
| South Africa | Sunflower | 215 |
| Zimbabwe | Jatropha | 100 |

The underutilization of agricultural potentials in many African countries has also contributed to the increased push by domestic interests towards finding alternative outlets for their productive capacity. Most developed countries are also moving from voluntary legislation to obligatory legislation imposing market share of biofuels in the transport sector. A mandatory blending legislation as well as internalization of biofuel external costs is required to reduce carbon dioxide emissions and improve energy security. To meet the European Union's aim of 5.75% market share of biofuels in 2010 and about 8% by 2020, countries will have to import feedstock from elsewhere, due to lack of sufficient arable land for energy crops and the well-established regulations safeguarding forests and governing land use. These EU biofuel targets have already triggered many investments in tropical biofuels [29]. The notion is that Africa has vast unused land areas that can be employed for biofuel production in order to generate export incomes, employment for rural people and smallholders, and profit for foreign investors. The rush towards biofuel production among foreign investors in Africa has created problems for governments to coordinate and guide such production. In many African countries, investors have started exploiting biofuel/biodiesel production without the existence of policies and regulatory framework for such productions. These could result in negative impacts on the environment such as deforestation, biodiversity loss, and land use problems. African countries are at the various stages of initiating commercial production of biofuels to capture the benefits of their value chain. This has led to large-scale mechanized feedstock/biofuel production, based on a model of industrial agriculture that uses large-scale pesticide and fertilizer input and measures productivity and effectiveness largely by the economic benefits with little consideration of social and environmental costs and benefits [30]. Sustainable development of biofuel industry in Africa can help kick-start investment in Africa's neglected agricultural sector, with positive spillover effects in terms of poverty reduction, food crop production and energy diversification, but governance and policy are required to ensure that these developments are carried out in a way that helps African countries achieve sustainable development. The concept of biofuels sustainability is extremely complex, and the views are diverse. To assess the sustainability of biofuels, the impact of the production, trade, and final conversion of the biofuel must be analyzed using an integrated approach and taking into account the three interlinked criteria of sustainable development such as social impacts (social well-being), environmental impacts (maintaining or improving environmental quality), and socioeconomic impacts (economic viability of biofuels production and associated welfare considerations) [31]. The fundamental issue with the production of biofuels is the use of landresources and the competition with food production. Therefore, agricultural practices need to be integrated; the Earth's natural resources, especially those of soil, water, plant and animal diversity, climate, and ecosystem services, are fundamental for the structure and function of agricultural systems in support of life on Earth. Historically, the developments in agriculture have concentrated on increasing productivity and effectively exploiting natural resources, but they have ignored complex interactions between agricultural activities, local ecosystems, and society. There are, therefore, several opportunities that can be realized through the integration of biofuel production with food production. There are also recent biofuels technological developments that can be applied to the usage of agricultural wastes for the production of biofuels [32].

# BIODIESEL PRODUCTION

Biodiesel is defined as a fatty acid alkyl derived from the transesterification of vegetable oils or animal fats [33]. It is the product obtained when a vegetable oil or animal fat (triglyceride) reacts with an alcohol in the presence of a catalyst of which glycerol is a byproduct. Biodiesel can be prepared from a variety of sources including vegetable oils such as oilseeds, animal fats, waste oils, and greases [34]. In light of making biodiesel production more practical economically, the possibilities of converting its glycerol byproduct into value-added chemicals will provide more economically viable alternatives for the biodiesel industry. This tendency may lead to a decrease in biodiesel prices and could improve the glycerol market. Studies have shown that the glycerol commodity market is very limited and any increase in biodiesel production will cause a sharp decline of more than 50% of its current value. With the intended increased expansion of biodiesel productions and consequent potential decrease of glycerol prices, glycerol is expected to become a major platform chemical and has been recently identified as an important building block for future bio-refineries thus adding credit and value to the glycerol byproduct [35]. As a result, new opportunities and challenges exist for research and development in the biodiesel industry to propel the synergy needed for the transformation of glycerol into alue-added chemicals. In South Africa, the medium-scale production-biodiesel ranges between 200,000 litres to 300,000 liters monthly with an output of about 40,000 liters of glycerol monthly [36]. The price of glycerol obtained from waste oil feedstock was about 50 cents/liter, while the price of that obtained from refined oil was about 2 dollars per liter. The former is mainly used as an additive in animal feed requiring a maximum of 3% in total feed, also predominantly in local soap industries, and to produce antifreeze. The remaining is either stored away or disposed.

# Value-Added Applications of Glycerol

Glycerol is a colorless, odorless, hygroscopic, and sweet-tasting viscous liquid, and because of its humectants properties, it is widely utilized in the cosmetic, pharmaceutical, and food industries; also has wide applications in lacquers, plastics, alkyl resins, tobacco, explosives, and cellulose processing. As reported by Yazdani and Gonzalez [37], the tremendous growth of the biodiesel industry has created a glycerol surplus that has resulted in a dramatic ten-fold decrease in crude glycerol prices; it is expected that the projected volume production of crude glycerol over the next five years will exceed the present commercial demand for purified glycerol, thus, new routes to convert crude glycerol, which is now considered a waste with an associated disposal cost, into higher-value products demand urgent attention.

The accumulation of crude glycerol resulting from biodiesel production propelled new ideas where some scientists tested the possibility of applying cheap crude glycerol as an animal feed ingredient in place of maize [38–41]. It has been found to be an attractive energy source for animal feed because of the similar energy value in comparison with corn and soybean meal; it, however, does pose a potential danger to the animals when the qualities of this biodiesel byproduct are not properly monitored [41]. The feedstock source and manufacturing process of biodiesel production are key factors in determining the composition of crude glycerol and its nutritional value [42]. It is, thus, necessary to analyze the physical, chemical, and nutritional properties of crude glycerol before including them into animal diet. A potential hazardous compound in crude glycerol is methanol with an acceptable level of methanol in crude glycerol to be used as a supplement in forage not exceeding 150 ppm as methanol poisoning may cause central nervous system injury, weakness, headache, vomits, blindness, or Parkinsonian-like motor diseases in animals [43]. Lammers et al. [40] report that forage can be supplemented with 15% of glycerol without negative influence on egg production. Schieck et al. [44] stated that glycerol up to 9% can be used as a supplement in diet of lactating sows as an alternative energy source instead of maize. Previous research on finding new applications of glycerol includes its use as a low-cost feedstock for functional derivatives and additives for concrete chemicals and as a precursor of valued fine chemicals [45]. Low-value agricultural use involves mixing crude glycerol with manure for fertilizer or with feed for animals. Studies conducted by Donkin and Doane [38] indicate that glycerol as a valuable feed ingredient can be included as a macroingredient in diets for lactating dairy cows without any poisonous effect, presenting an alternative strategy for the formulation of diets when maize is not readily available. Glycerol can also be potentially transformed via bacteriologic processes into products that can be used for plastic production [46]. A promising and equally economically advantageous use is the conversion of glycerol into high-value products such as mono- and difatty acids [47]. Focusing on recent developments in the conversion of glycerol into value-added chemicals in future biorefineries, is the use of glycerol or its derivatives as a potential gasoline additive especially in the industrial synthesis of GTBE (glycerol tert-butyl ether) as shown in Figure 2. Methyl tertiary-butyl ether (MTBE) used as an anti-knocking agent has been discredited due to its carcinogenity and its tendency to contaminate ground water [17]; GTBE, on the other hand, is considered safer than MTBE because it does not readily mix with water and, hence, reduces the likelihood of contaminating ground water in case of a spill [45]. Glycerol cannot be added directly to fuel because at high temperatures it polymerizes and, therefore, clogs the engine while partly getting oxidized to toxic acrolein. In this respect, glycerol tertiary-butyl ether (GTBE) poses as an excellent additive with a large potential for diesel and biodiesel reformulation. Various acetals and ketals can also be used as ignition accelerators and antiknock additives in combustion engines, proved to have lower levels of particle emissions, and can be added up to 10% volume of the fuels used, they can also be used as the base for surfactants [48]. According to a study [49], different oxygenated glycerol derivatives were synthesized from glycerol including acetals and ketals, and their emissions were evaluated with different kinds of vehicles and engines, and it was found out that a blend with 10% GTBE could be interesting to reduce particulate matter emissions by more than 20%. Stating from an economical point of view, the manufacturing cost of GTBE is low among the different derivatives, and GTBE is a very promising additive, and the benefits of blending GTBE in diesel fuel not limited to its performance for reducing particulate matter (PM) emissions; it also plays a part in the reduction of $CO_2$ emissions. Various tests carried out by Jaecker-Voirol et al. [49] deduce that the use of this reformulation does not present any technical disadvantage; therefore, GTBE could be considered as an excellent additive with a large potential for diesel and biodiesel reformulation. The use of GTBE in Africa would have significant health benefits in replacing lead as an octane enhancer in most African countries where leaded fuel is still widely used; of a total of 49 countries in the Subsaharan Africa, 22 countries only use leaded fuel; only 14 use a dual system, while 13 use the unleaded fuel [50]. The ultimate decision, however, to use GTBE in biodiesel formulation would be controlled by economic criteria and the willingness on the part of industry. In other perspectives, the use of anaerobic fermentation to convert abundant and low-priced glycerol streams generated in the production of biodiesel into higher-value products also presents a promising route to attain economic viability in the biofuels industry. The use of crude glycerol has also been verified as an ideal carbon source for the fermentation processes by microorganisms [51]. Clearly, the development of processes to convert crude glycerol into higher-value products as shown in Figure 1 is both an urgent need and a "target of opportunity" for the development of biorefineries. Such technologies could be readily integrated into existing biodiesel facilities, thus establishing true biorefineries and revolutionizing the biodiesel industry by dramatically improving its economics [47]. In terms of green chemistry, the incorporation of glycerol into biofuel can improve the efficiency of the process without substantial modifications of its physicochemical properties with a novel biofuel patent denoted as Ecodiesel-100 obtained via 1, 3-selective partial ethanolysis of triglycerides with a lipase. Studies conducted reveal that the Ecodiesel-100 incorporates glycerol in its mono-phasic homogeneous mixtures requiring no additional separation steps as compared with conventional biodiesel production while proffering the advantage of the elimination of byproducts during the preparation process [52]. Recent studies have also demonstrated that the presence of mono triglyceride adds value to the biofuel by improving the lubricity on the engine and the atom efficiency is also improved as the total number of atoms involved in the reaction is part of the final mixture. The possibility of using crude glycerol byproduct from the biodiesel industry as a carbon source for microalgae that produces omega-3 fatty acids has also been established, adding to the fact that impurities in crude glycerol may actually be beneficial to algal growth [53].

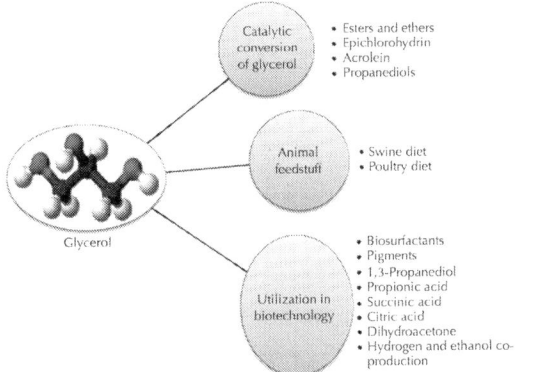

**Figure 1:** Value-added chemicals obtainable from glycerol.

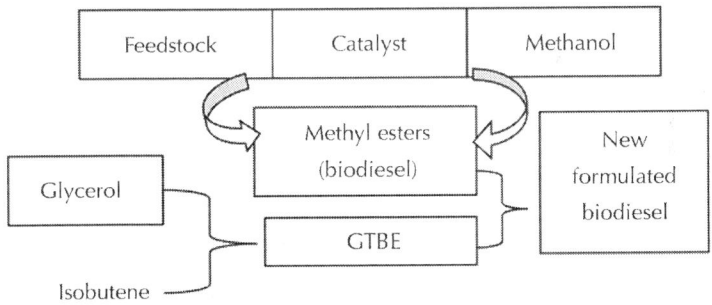

**Figure 2:** Biodiesel formulation using GTBE..

# CONCLUSIONS

Biofuels have a crucial role to play in fulfilling the long-term goal of limiting the growth of GHG emissions and the transition of the current petroleum-based society towards a more sustainable one. Biodiesel production could offer the developing African countries potential economic, ecological, social, and security benefits while pursuing and advocating the implementation of biodiesel use on the continent.

Surely, there is a large room for improvement in biodiesel production; strong political will on the part of governments will go a long way in achieving a prospect of self-reliance with respect to energy supplies at national and local levels. However, the cost of biodiesel production is quite high, and it exceeds prices of traditional; fuels, thus the inevitability of making biodiesel production more practical economically via the possibilities of converting its glycerol byproduct into value-added chemicals will provide more economically viable alternatives for the biodiesel industry. The utilization of crude glycerol into higher-value products would greatly increase the net energy, sustainability and profitability of the biodiesel industry overall. In summary, the pace of biodiesel sustainability can be met on the African continent if all hands of producers, research, and development in educational institutions and governments (politicians and political structures) are on deck

# ACKNOWLEDGMENTS

The authors wish to acknowledge the contributions of late Dr. Bamikole Amigun to this paper.

# REFERENCES

1.  W. Tsai, "Energy sustainability from analysis of sustainable development indicators: a case study in Taiwan,"Renewable and Sustainable Energy Reviews, vol. 14, no. 7, pp. 2131–2138, 2010.

2.  B. Amigun, F. Müller-Langer, and H. von Blottnitz, "Predicting the costs of biodiesel production in Africa: learning from Germany,"Energy for Sustainable Development, vol. 12, no. 1, pp. 5–21, 2008.

3.  K. J. Singh and S. S. Sooch, "Comparative study of economics of different models of family size biogas plants for state of Punjab, India," Energy Conversion and Management, vol. 45, no. 9-10, pp. 1329–1341, 2004.

4.  B. Amigun, J. K. Musango, and W. Stafford, "Biofuels and sustainability in Africa," Renewable and Sustainable Energy Reviews, vol. 15, no. 2, pp. 1360–1372, 2011.

5. A. I. Eleri and E. O. Eleri, "Prospects for Africa-Rethinking Biomass Energy in Sub-Sahara Africa. Prospects for Africa-Europe Policies," The VENRO -Project on the Africa -Eu Partnership, Bonn, Germany, 2009.

6. W. K. Biswas, P. Bryce, and M. Diesendorf, "Model for empowering rural poor through renewable energy technologies in Bangladesh," Environmental Science and Policy, vol. 4, no. 6, pp. 333–344, 2001.

7. S. Fernando, S. Adhikari, K. Kota, and R. Bandi, "Glycerol based automotive fuels from future biorefineries," Fuel, vol. 86, no. 17-18, pp. 2806–2809, 2007.

8. R. S. Karinen and A. O. I. Krause, "New biocomponents from glycerol," Applied Catalysis A, vol. 306, pp. 128–133, 2006.

9. T. Willke and K.-D. Vorlop, "Industrial bioconversion of renewable resources as an alternative to conventional chemistry," Applied Microbiology and Biotechnology, vol. 66, no. 2, pp. 131–142, 2004.

10. O. Davidson and M. Chenene, Sustainable Energy in Sub-Saharan Africa, ICSU Regional Office For Africa Science Plan Report Pretoria, International Council for Science, Johannesburg, South Africa, 2007.

11. R. Blinca, A. Zidans ka, and I. Šlausc, "Sustainable development and global security," Energy, vol. 32, pp. 883–890, 2007.

12. V. Schalkwy, http://www.allafrica.com/, 2012.

13. M. Balat and H. Balat, "A critical review of bio-diesel as a vehicular fuel," Energy Conversion and Management, vol. 49, no. 10, pp. 2727–2741, 2008.

14. W. Thurmond, "Global Biodiesel Market trends, Outlook and opportunities (pdf)," http://www.unece.isu.edu/biofuels/documents/2010Aug/b10_16.pdf, 2012.

15. M. Johnston and T. Holloway, "A global comparison of national biodiesel production potentials," Environmental Science and Technology, vol. 41, no. 23, pp. 7967–7973, 2007.

16. W. Thurmond, "Biodiesel 2020: A global Market survey, Volume two," Emerging markets, http://www.emerging-markets.com/, 2012.

17. B. Amigun, R. Sigamoney, and H. von Blottnitz, "Commercialisation of biofuel industry in Africa: a review," Renewable and Sustainable Energy Reviews, vol. 12, no. 3, pp. 690–711, 2008.

18. "Africa's biofuel boom: Mozambique," http://www.iol.co.za/index.php/,http://www.worldwatch.org/node/5450, 2012.

19. A. A. Kiss, A. C. Dimian, and G. Rothenberg, "Solid acid catalysts for biodiesel production-towards sustainable energy," Advanced Synthesis and Catalysis, vol. 348, no. 1-2, pp. 75–81, 2006.

20. "Journey to Forever. National standards for biodiesel,"http://www.journeytoforever.org/biodiesel_yield2.html#biodstds, 2006.

21. S. Mwakasonda, "Moving toward sustainable biofuel programs in Africa," in Proceedings of the 1st High-Level Biofuel Seminar in Africa, Addis Ababa, Ethiopia, 2007.

22. M. Canakci and J. Van Gerpen, "A pilot plant to produce biodiesel from high free fatty acid feedstocks,"Transactions of the American Society of Agricultural Engineers, vol. 46, no. 4, pp. 945–954, 2003.

23. L. Chapman, "Transport and climate change: a review," Journal of Transport Geography, vol. 15, pp. 354–367, 2007.

24. M. Lenzen, C. Dey, and C. Hamilton, "Climate change," in Handbook of Transport and the Environment, pp. 37–60, Elsevier, Amsterdam, The Netherlands, 4th edition, 2003.

25. M. Nolte, Commercial biodiesel production in South Africa: a preliminary economic feasibility study [M.S. thesis], Department of Chemical Engineering, University of Stellenbosch, Stellenbosch, South Africa, 2007.

26. B. J. Waterson, B. Rajbhandari, and N. B. Hounsell, "Simulating the impacts of strong bus priority measures," Journal of Transportation Engineering, vol. 129, no. 6, pp. 642–647, 2003.

27. M. C. Monfort, "Global trends in seafood and sustainability: market movements and trends," inProceedings of the Seafood Choice Alliance Conference, Barcelona, Spain, 2008.

28. IEA, "World energy outlook 2007," International Energy Agency Parishttp://www.mybiodiesel.com/biodiesel-history.php, 2007.

29. G. Sorda, M. Banse, and C. Kemfert, "An overview of biofuel policies across the world," Energy Policy, vol. 38, no. 11, pp. 6977–6988, 2010.

30. H. S. Dillon, T. Laan, and H. S. Dillon, "Biofuels-at what cost? Government support for ethanol biodiesel in Indonesia. The Global Studies Initiative," Part of the International Institute for Sustainable Development, 2008.

31. E. Smeets, M. Junginger, A. Faaij, A. Walter, P. Dolzan, and W. Turkenburg, "The sustainability of Brazilian ethanol-An assessment of the possibilities of certified production," Biomass and Bioenergy, vol. 32, no. 8, pp. 781–813, 2008.

32. S. R. Gliessman, Agroecology: Ecological Processes in Sustainable Agriculture, Ann Arbor Press, Michigan. Mich, USA, 1998.

33. M. S. Friedrich, A worldwide review of the commercial production of biodiesel-a Technological, economic and ecological investigation based on case studies [M.S. thesis], Institute fur Technologie und Nachhaltiges produkt management, Vienna, Austria, 2004.

34. D. Dermirbas and S. Karslıoglu, "Biodiesel production facilities from vegtable oils and animal fats,"Energy Source, vol. 29, pp. 33–141, 2007.

35. K. S. Tyson, J. Bozell, R. Wallace, E. Petersen, and L. Moens, "Biomass oil analysis," Research Needs and Recommendations NREL/TP-510-34796, 2004.

36. O. Nieuwoudt, "Glycerol inventory in South Africa," Private Communication, 2011.

37. S. S. Yazdani and R. Gonzalez, "Anaerobic fermentation of glycerol: a path to economic viability for the biofuels industry," Current Opinion in Biotechnology, vol. 18, no. 3, pp. 213–219, 2007.

38. S. S. Donkin and P. H. Doane, "Glycerol from biodiesel production: the new corn for dairy cattle,"Nutrition Research, vol. 37, 2008.

39. S. Cerrate, F. Yan, Z. Wang, C. Coto, P. Sacakli, and P. W. Waldroup, "Evaluation of glycerine from biodiesel production as a feed ingredient for broilers," International Journal of Poultry Science, vol. 5, no. 11, pp. 1001–1007, 2006.

40. P. J. Lammers, B. J. Kerr, T. E. Weber et al., "Digestible and metabolizable energy of crude glycerol for growing pigs," Journal of Animal Science, vol. 86, no. 3, pp. 602–608, 2008.

41. W. A. Dozier, B. J. Kerr, A. Corzo, M. T. Kidd, T. E. Weber, and K. Bregendahl, "Apparent metabolizable energy of glycerin for broiler chickens," Poultry Science, vol. 87, pp. 317–322, 2008.

42. C. F. Hansen, A. Hernandez, B. P. Mullan et al., "A chemical analysis of samples of crude glycerol from the production of biodiesel in Australia, and the effects of feeding crude glycerol to growing-finishing pigs on performance, plasma metabolites and meat quality at slaughter," Animal Production Science, vol. 49, no. 2, pp. 154–161, 2009.

43. D. C. Dorman, J. A. Dye, M. P. Nassise, J. Ekuta, B. Bolon, and M. A. Medinsky, "Acute methanol toxicity in minipigs," Fundamental and Applied Toxicology, vol. 20, no. 3, pp. 341–347, 1993.

44. S. J. Schieck, B. J. Kerr, S. K. Baidoo, G. C. Shurson, and L. J. Johnston, "Use of crude glycerol, a biodiesel coproduct, in diets for lactating sows," Journal of Animal Science, vol. 88, no. 8, pp. 2648–2656, 2010.

45. A. Behr, J. Eilting, K. Irawadi, J. Leschinski, and F. Lindner, "Improved utilisation of renewable resources: new important derivatives of glycerol," Green Chemistry, vol. 10, no. 1, pp. 13–30, 2008.

46. B. A. Hunt, "Production of Ethers of Glycerol from Crude Glycerol-The By-Product of Biodiesel Production," http://digitalcommons.unl.edu/chemeng, 1998.

47. C. Zhou, J. N. Beltramini, Y. Fan, and G. Q. Lu, "Chemoselective catalytic conversion of glycerol as a biorenewable source to valuable commodity chemicals," Chemical Society Reviews, vol. 37, no. 3, pp. 527–549, 2008.

48. A. Piasecki, A. Sokolowski, and B. Burczyket, "Fatty Acid condensates," Journal of the American Oil Chemists› Society, vol. 74, pp. 33–37, 1997.

49. A. Jaecker-Voirol, I. Durand, G. Hillion, B. Delfort, and X. Montagne, "Glycerin for new biodiesel formulation," Oil and Gas Science and Technology, vol. 63, no. 4, pp. 395–404, 2008.

50. V. Thomas and A. Kwong, "Ethanol as a lead replacement: phasing out leaded gasoline in Africa,"Energy Policy, vol. 29, no. 13, pp. 1133–1143, 2001.

51. N. Özbay, N. Oktar, and N. A. Tapan, "Esterification of free fatty acids in waste cooking oils (WCO): role of ion-exchange resins," Fuel, vol. 87, no. 10-11, pp. 1789–1798, 2008.

52. R. Luque, L. Herrero-Davila, J. M. Campelo et al., "Biofuels: a technological perspective," Energy and Environmental Science, vol. 1, no. 5, pp. 542–564, 2008.

53. Z. Wen, "Researcher converts biodiesel-waste glycerol into omega-3 fatty acids,"http://www.phys.org/news138543542.html, 2012.

# Chapter 7

# Chitosan and its Derivatives Applied in Harvesting Microalgae for Biodiesel Production: An Outlook

Guanyi Chen[1] Liu Zhao,[1] Yun Qi,[1] and Yuan-Lu Cui[2]

[1]School of Environment Science and Engineering, State Key Laboratory of Engines, Tianjin University, No. 92 Weijin Road, Nankai District, Tianjin 300072, China
[2]Tianjin State Key Laboratory of Modern Chinese Medicine, Tianjin University of Traditional Chinese Medicine, Tianjin 300193, China

## ABSTRACT

Although oil-accumulating microalgae are a promising feedstock for biodiesel production, large-scale biodiesel production is not yet economically feasible. As harvesting accounts for an important part of total production cost, mass production of microalgae biodiesel requires an efficient low-energy harvesting strategy so as to make biodiesel production economically attractive. Chitosan has emerged as a favorable flocculating agent in harvesting of microalgae. The aim of this paper is to review current research on the application of chitosan and chitosan-derived materials for harvesting microalgae. This offers a starting point for future studies able to invalidate, confirm, or complete the actual findings and to improve knowledge in this field.

# INTRODUCTION

Fossil fuels currently account for about 80% of global energy production. As demand for energy continues to increase, oil prices will rise and fossil fuels will be exhausted at some point in the future. Meanwhile, extensive utilization of fossil fuels has led to adverse effects including global climate change, environmental pollution, and public health problems [1].

Therefore, many countries have started to take series of measures to resolve this problem [2]. Identifying alternative renewable sources of fuel that are carbon neutral is important for many countries. Among the potential sources of renewable energy, biodiesel is of particular interest. Major advantages of biodiesel include mitigation of carbon dioxide emissions and potential use as a substitute for petroleum [3]. Biodiesel also has a higher energy density than competing biofuels measured in kilojoules per unit of mass. Furthermore, widespread adoption of biodiesel could improve urban air quality as emissions of carbon monoxide and volatile organic compounds are significantly lower than petroleum-derived diesel. Biodiesel may become a primary energy source for sustainable development and could play a crucial role in the global energy infrastructure in the future. Energy security may drive biodiesel production as nations attempt to reduce their reliance on petroleum imports.

More than 95% of biodiesel sources are first generation agricultural edible crop oils [4] such as palm oil, oilseed rape, and soybean. However, these vegetable oils are also used for human consumption, which may lead to an increase in price of food-grade oils. Meanwhile, any transition to biodiesel production based upon these crops would require arable land, with a corresponding drop in other forms of agricultural productivity. In theory, it should be possible to produce low-cost biodiesel using nonedible oils (second generation biofuels), such as frying oils, animal fats, soap stocks, and grease. However, these nonedible oils are rarely available in quantities suitable for industrial-scale biodiesel production [5]. As a consequence, there is renewed interest in methods of producing biodiesel from microalgae.

As a promising feedstock for biodiesel production, microalgae offers compelling advantages compared to other oil crops: (1) the cultivation of microalgae does not need much land as terraneous plants [6]; indeed, biomass growth might not require any arable land at all if offshore farming proves feasible [7]; (2) microalgae have much higher biomass productivities than land plants; (3) some microalgae species are rich in oils; they can accumulate up to 20–50% (w/w DW) triacylglycerols [8] and certain strains may have content as high as 85% lipid under limited condition [9]; (4) microalgae utilize $CO_2$ from the atmosphere via photosynthesis, offsetting greenhouse gas emissions; (5) microalgae require less freshwater for cultivation than terrestrial crops. Microalgae growth effectively removes nutrients, such as nitrogen and phosphorus, and heavy metals from wastewater; (6) biodiesel produced from microalgal oil has advantageous properties compared to standard biodiesel. These advances suggest that industrial production of biodiesel from microalgal oils may be feasible in the near future.

# Microalgae Harvest Strategy

Although oil-accumulating microalgae are a promising feedstock for biodiesel production, large-scale biodiesel production is not yet economically feasible. This is mainly due to the high-energy inputs required for harvesting [10]. However, as microalgae' diameters are often as small as 3–30 mm, harvesting the microalgae is a significant problem. In some commercial production systems, the culture broths are below $0.5$ kg/m$^3$ dry biomass, which means that huge volumes need to be handed before algae oil can be extracted. Molina et al. estimated that harvesting can account for 20–30% of the total production cost [11]. Chisti even reported that the cost of the recovery process in his study contributed about 50% to the final cost of oil production [6].

Consequently, mass production of microalgae biodiesel acquires efficient low-energy harvesting strategy so as to make biodiesel production economically feasible. Microalgae can be harvested by centrifugation, filtration, flotation, sedimentation, and electrophoresis techniques [12].

Centrifugation can recover most microalgae from the liquid broth and is used in many commercial systems. Although centrifugation is effective, this process is energy intensive [13], which reduces the net energy return on investment (EROI) from the biodiesel produced from the microalgae, making this option less attractive both in financial and environmental terms. Norsker et al. calculated centrifugation required as much as 50% of the available energy in the recovered biomass [14]. From an energetic point of view, harvesting a large amount of microalgae using centrifugation is time consuming and costly [15]. Filtration is another option for harvesting cells, but this technology is only useful for the harvest of large species such as Spirulina but fails to recover small microalgae such as Chlorella or Scenedesmus [10].

Coagulation/flocculation processes offer high microalgae biomass recovery at reasonable costs [16]. These harvesting techniques have been successfully used in aquaculture, wastewater treatment, and removal of microalgae [10] and also reduce the net energy input required to produce biodiesel from microalgae feedstock.

Several flocculants have been developed to induce flocculation of microalgae cells that can be applied to the treatment of large amount of microalgae. According to their chemical compositions, there are two classifications of flocculants: inorganic flocculants and organic flocculants/polyelectrolyte flocculants [1].

The commonly used inorganic flocculants include ferric chloride $(FeCl_3)$, aluminum sulfate $(Al_2(SO_4)_3)$, and ferric sulfate $(Fe_2(SO_4)_3)$ [17]. These multivalent metal salts are effective flocculants or coagulants and have been widely used to flocculate algal biomass. prepolymerized metal salts (such as polyaluminium chloride and polyferric sulfate) have proved to be efficient over a wider pH range [18]. However, flocculation by metal salts is not an appropriate method for cheap and sustainable harvesting of microalgae in large-scale microalgae culture. This is because these flocculants are expensive and may produce large amounts of sludge, which can kill or prevent the growth of the microalgae and leave a residue in the water, and the excess cationic flocculant needs to be removed before it can be reused [19]. Another disadvantage of metal salts as flocculants is the concern about the human health, such as involvement in Alzheimer's disease and carcinogenesis [20–22].

The organic flocculants, such as synthetic polymeric flocculants and modified natural polymers, have also been applied to the microalgae harvesting process. In the last few decades, chitosan has emerged as a favorable flocculating agent in harvesting microalgae. Chitosan is becoming increasingly important as a natural biopolymer due to its unique combination of properties like biodegradability, biocompatibility, renewability, bioactivity, and ecological acceptability, in addition to attractive physical and mechanical properties [23]. It also has variety of current and potential applications in wastewater treatment [24], biomedical engineering [25], food processing [26], and so forth. In particular, chitosan has also been examined to formulate nanoparticles to facilitate targeting drug to specified organ [27].

# GENERAL ASPECTS OF CHITOSAN

Chitosan, poly-$\beta$(1-4)-2-amino-2-deoxy-d-glucopyranose, is a cationic polyelectrolyte obtained by deacetylation of chitin. Chitin and chitosan are both aminoglucopyrans composed of N-acetylglucosamine (GlcNAc) and glucosamine (GlcN) residues. The chemical structures of chitin and chitosan are shown in Figure 1 [28]. Acetamide group of chitin can be converted into amino group to give chitosan, when the degree of deacetylation of chitin reaches about 50% (depending on the origin of the polymer). Chitosan is also polyelectrolyte which has several million Daltons. The average molecular weight of commercially available chitosan ranges between 3800 and 20,000 Daltons and is 66% to 95% deacetylated. The chitosan toxicity tests showed that the toxicity was negligible [29]. Because chitosan can be produced in various forms such as powder, paste, film, and fiber, it is more widely applied in industry than chitin [30].

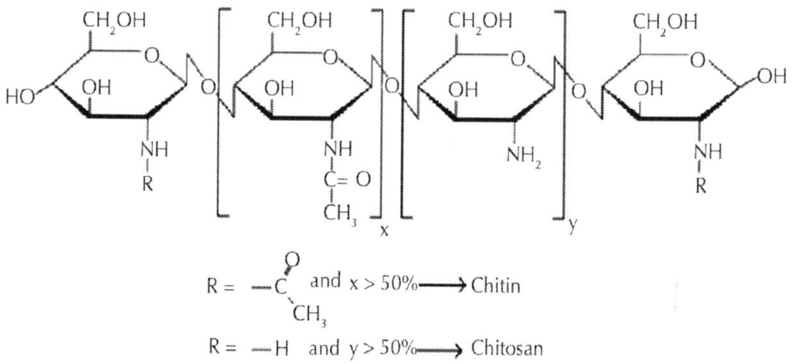

R = $-\overset{\displaystyle O}{\underset{\displaystyle CH_3}{C}}$ and x > 50% ⟶ Chitin

R = $-H$ and y > 50% ⟶ Chitosan

**Figure 1:** Structures of chitin and chitosan [28].

Due to the presence of three different polar functional groups ($-OH$, $-NH_2$, and C–O–C), chitosan has high water capacity [28]. Chitosan has the special quality of gelling upon contact with anions, forming beads under very mild conditions [31]. And the presence of amino groups makes chitosan a cationic polyelectrolyte (pKa = 6.5), one of the few found in nature [32]. Ordinary chitosan is insoluble in water at near neutral pH and most common organic solvents (e.g., DMSO, DMF, NMP, organic alcohols, and pyridine), which is attributed to extensive intramolecular and intermolecular hydrogen bonding between the chains and sheets, respectively [33]. Chitosan can be soluble in some diluted organic acids (such as acetic, formic, and lactic acids) and some inorganic acid. Although the distribution of acetyl groups along the chain may modify solubility [34], the solubilization is mainly due to the protonation of $NH_2$ groups on the $C_2$ position of the β-glucosamine unit.

# CHITOSAN APPLIED IN HARVESTING MICROALGAE

Chitosan not only has been proved highly effective for water treatment and environmental protection, but also shows interesting properties in removing both freshwater algae and marine algae.

The most likely mechanisms involved in this coagulation are adsorption and charge neutralization. Chitosan has a net positive charge because of the high charge density of the chitosan. As the overall charge of microalgae cells is negative, the positively charged chitosan is strongly adsorbed on microalgae cells, which results in most of the charged groups being close to the surface of the cells [35] and effectively destabilize the microalgae [36]. Chitosan first neutralizes charges on the microalgae cells, weakens the electrostatic repulsion between the microalgae cells, and then reduces the interparticle repulsion. Such effect is called charge neutralization [37].

Some authors report chitosan as an effective algal flocculant. Figure 2 shows the scan electron microscope (SEM) of Chlorella vulgaris before and after flocculation by chitosan. After flocculation, C. vulgaris algae were surrounded by chitosan, and the surface became fibrillar. This change of surface was observably caused by chitosan, which possesses heterogeneous surface structure [38].

**Figure 2:** (a) SEM of C. vulgaris before harvesting. (b) SEM of C. vulgaris after harvesting [38].

Using life cycle assessment (LCA), Beach compared the chitosan method to centrifugation and filtration/chamber press methods [39]. Figure 3 showed the system used as a basis for the comparison, where the cultivation and downstream were assumed to be equivalent among all methods. LCA showed that flocculation by chitosan for harvesting N. oleoabundans is the least energy intensive and had the best profile across all other categories of environmental impacts.

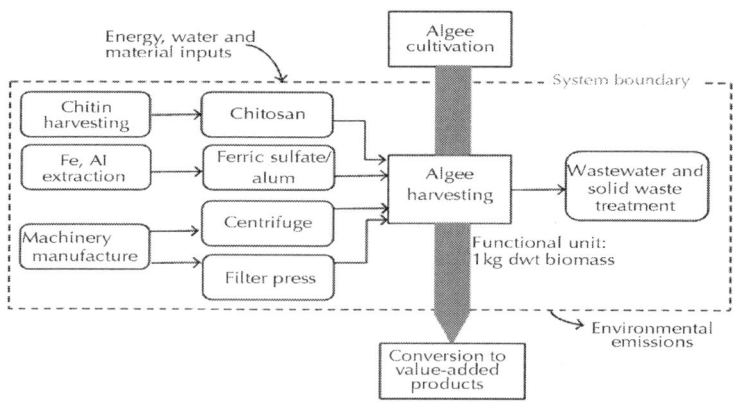

**Figure 3:** Overview of the system used as a basis for the comparison [39].

# Effect of Chitosan Dosage on Flocculation Efficiency

In general, chitosan can effectively flocculate algal species at 5 mg/L to 200 mg/L. Divakaran and Sivasankara Pillai [40] reported that chitosan successfully removed 90% of turbidity with chitosan concentration of 5 mg/L. Ahmad et al. [41] reported a % of Chlorella sp. removal at 10 ppm of chitosan. In order to obtain 95–100% flocculation efficiency, chitosan concentration was about 20 mg/L for Chlorella and 2 mg/L for S.Costatum [16]. Chitosan is required in low dosage in freshwater but its flocculating power is reduced in salt water.

Chitosan's ability to remove microalgae effectively at low dosage is partly caused by its' properties. Chitosan not only acts as an adsorbent, but also spontaneously coagulates to agglomerate the microalgae cells. However, when using overdose of chitosan, the percentage of microalgae cells removed declined sharply, which may be caused by charge neutralization and bridging phenomena. During the flocculation, cationic charge of chitosan attracts the negatively charged microalgae, reducing the electrostatic repulsion among microalgal cells and then forming the flocs. Excess amino group led to restabilization of the microalgae and decreasing of separation efficiency [38].

# Effect of PH on Flocculation Efficiency

The importance of pH on flocculation of microalgae was investigated by many researchers [41–44]. The influence of pH on chitosan's molecular structure can be due to differences in the protonation of the biopolymer amine groups and variations in the conformation of the macromolecule chain and in the structure of the flocs [37]. In alkaline solutions, the positive charge gradually disappeared and chitosan is able to produce large and dense flocks. When the pH increases to neutralization point, the algal cells have the highest negative charge, and the flocculation efficiency is enhanced as the electrostatic interaction between the algal cells and chitosan, whereas, in acidic solutions, chitosan becomes a more extended chain and therefore produces smaller looser flocs [20]. In the study of Divakaran and Sivasankara Pillai [40], maximum clarification was obtained at pH 7 for the freshwater species. Cheng et al. reported that a higher pH at

8.5 was optimal for Chlorella sorokiniana [42]. In the study of an even higher pH at 9.9, 90% of Phaeodactylum tricornutum was harvested with 20 ppm chitosan [18].

Morales suggested that chitosan activity and flocculation efficiency were increased as the viscosity and the mean surface charge of algal cells were decreased when pH was below 7 [16]. While Xu et al. reported that when the working pH was at 6 or 5, half the amount of chitosan was needed to induce effective flocculation compared to pH 7 [43].

These differences in response to pH can be explained by difference in culture media, growth conditions, and unique strain properties, such as cell morphology, extracellular organic matter, and cell surface charge [44].

# Effect of Algae Species on Flocculation Efficiency

The flocculation efficiency of algal suspension is different from one algal species to another. Chlorella vulgarisreached 99.7% removal rate at 200 mg/L chitosan [45]. While 150 mg/L of chitosan was required for optimal flocculation of Chaetoceros muelleri [46]. For Skeletonema costatum 2 mg/L of chitosan was needed when 95% flocculation efficiency was obtained. Optimal flocculation of some species was list in Table 1.

Table 1: Optimal flocculation of some microalgae

| Species | Chitosan dosage | Parameters | Flocculation efficiency | References |
|---|---|---|---|---|
| Chaetoceros calcitrans | 80 mg/L | pH 8.0 | 80% | [46] |
| Chaetoceros muelleri | 150 mg/L | pH 8.0 | 95% | [46] |
| Chlorella | 5.0 mg/L | pH 7.0, | 90% | [63] |
| Chlorella consortium | 25 mg/L | Stirring for 1 min, settling for 10 min | 58±8 | [10] |
| Chlorella sorokiniana | 10 mg/gram algal dry weight | pH 6 | 99% | [43] |

| Chlorella sorokiniana | 25 mg/L | Stirring for 1 min, settling for 10 min | 30±11 | [10] |
|---|---|---|---|---|
| Chlorella sp. | 10 ppm | Mixing for 20 min, mixing at 150 ppm, and sedimenting for 20 min | 99.00±0.4% | [41] |
| Chlorella vulgaris | 200 mg/L | In a logarithmic growth phase | 99.7% | [45] |
| Chlorella vulgaris | 30 mg/L | pH 8.7, 300 rpm, and settling for 10 min. | 92% | [38] |
| Chlorella vulgaris and Microcystis sp. | 214 mg/L | Fish-processing wastewater, agitation speed of 131 rpm | 91.9% | [64] |
| Chlorococcum sp. | 25 mg/L | Stirring for 1 min, settling for 10 min | 38±11 | [10] |
| Nannochloropsis sp. | 100 mg/L | pH 9.0 | 92% | [62] |
| Neochloris oleoabundans | 100 mg/L | Mixing time for 10 min, mixing rate of 350 rpm | 95% | [39] |
| Pavlova lutheri | 80 mg/L | pH 8.0 | 80% | [46] |
| Phaeodactylum tricornutum | 20 mg/L | pH 9.9 Settling time of 10 min | 92% | [18] |
| Scenedesmus costatum | 2 mg/L | pH 7 | 95–100% | [16] |
| Scenedesmus obliquus | 25 mg/L | Stirring for 1 min, settling for 10 min | 20±15 | [10] |
| Scenedesmus quadricauda | 10 mg/L | pH 8.0; SDS was used as the collector | 90% | [65] |
| Skeletonema costatum | 80 mg/L | pH 8.0 | 70% | [46] |

| Synechocystis | 15 mg/L | pH 7.0 | >90% | [63] |
|---|---|---|---|---|
| Tahitian Isochrysis | 40 mg/L | pH 8.0 | 90% | [46] |
| Tetraselmis chui | 40 mg/L | pH 8.0 | 80% | [46] |
| Thalassiosira pseudonana | 40 mg/L | pH 8.0 | 90% | [46] |

# CHITOSAN MODIFIED FLOCCULANTS AND THEIR FLOCCULATION EFFICIENCY

Chitosan exhibits limitations in its reactivity and process ability. For a breakthrough in utilization of chitosan in flocculation of microalgae, chitosan modification to introduce a variety of functional groups will be a key point. Chitosan can be modified by chemical or physical processes to improve the mechanical and chemical properties. The efficiency of adsorption depends on physicochemical properties, mainly surface area, porosity, and particle size of adsorbents. As such procedure would not change the fundamental skeleton of polymers, chitosan would keep its original physicochemical and biochemical properties and finally would bring new or improved properties [47]. A great number of chitosan derivatives have been obtained by grafting new functional groups on the chitosan backbone to increase adsorption capacity. The new functional groups are incorporated to increase the density of sorption sites, to change the pH range for sorption, and to change the sorption sites in order to increase sorption. The chemical modification affords a wide range of derivatives with modified properties for specific use and applications in diversified areas mainly of pharmaceutical, biomedical, and biotechnological fields [32]. Such modification can also be applied to harvesting microalgae.

## Chitosan Modified Soils

Chitosan modified by soil particles (including the silica sand and local soil) showed highly effective in flocculating algae cells [48–52]. Pan and coworkers found that local soil particles including sand were critical for speeding up the kinetic processes of flocculation and sedimentation of algal flocs. The polymeric netting and bridging function of chitosan were the key mechanisms that allowed local soil particles to quickly flocculate algal biomass. Chitosan modified adsorbent has the functions of both flocculation and adsorption. Figure 4(a) shows the function of bridging, while Figure 4(b) illustrates the function of adsorption [53]. The chitosan made a "net" that captured the algae cells and other particles, and the soils provided the ballast or mass to carry the aggregates to the bottom. Chitosan was also important in inhibiting the escape of cells from the flocs. Chitosan and polyaluminium chloride used together as modifiers make it possible to use local beach sand for harvesting microalgae in seawater [50].

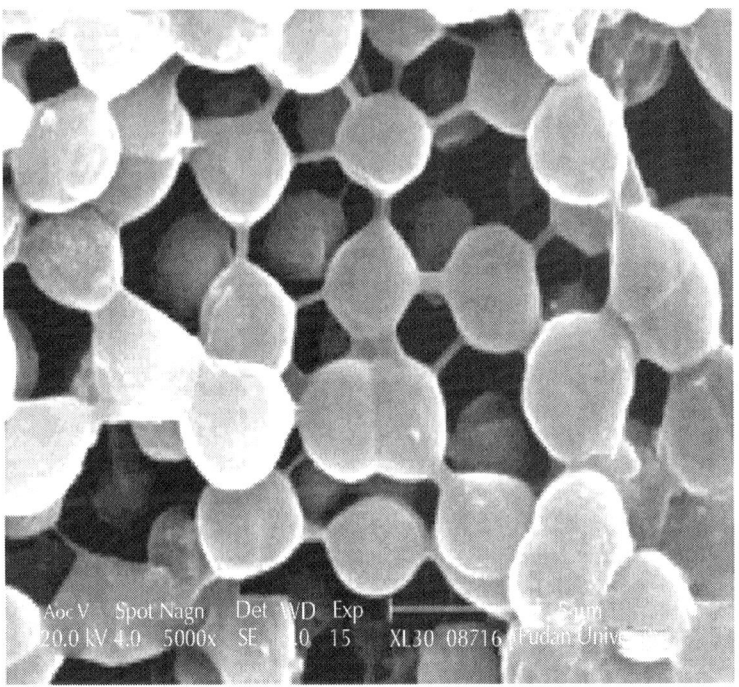

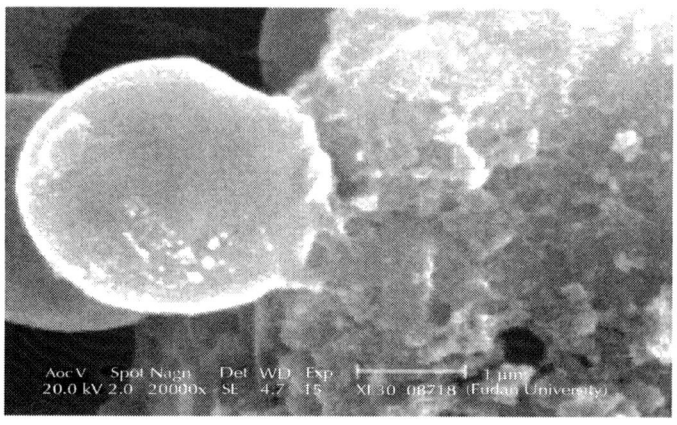

**Figure 4:** SEM images of algae treated with chitosan modified adsorbent [53].

Shao et al. studied the physiological responses of Microcystis aeruginosa under the stress of chitosan modified kaolinite (CMK). When flocculated with CMK, Chl a, carotenoids, phycocyanin, and allophycocyanin were much lower than the control. The results indicated that high level of CMK could cause cellular membranes damage and then the intracellular substances leakage and finally could cause the death of M. aeruginosa NIES-843 cell. Figure 5 showed that the strain can be effectively flocculated by CMK at 80 and 160 mg/L. However, the Microcystis cultures turned to be bluish at that time, which indicated the leakage of phycobilins [54].

**Figure 5:** Flocculation character of CMK and M. aeruginosa NIES-843 at different CMK loading levels [54].

# Aluminum Chloride and Aluminium Sulphate Modified Chitosan Used as Flocculants

## *Aluminum Chloride Modified Chitosan*

Chitosan could enhance the flocculation performance of polyaluminium chloride (PAC), when the high algae-laden water was treated by coagulation/flocculation/dissolved air flotation (C/F/DAF) [55]. The removal rate of algae cells was increased apparently compared with adding PAC alone. The structure and strength of flocs were improved when less than 1.0 mg/L chitosan was added, which significantly reduced the residual aluminum concentration.

Zhang et al. reported a composite coagulant (PACl-CTS), which was made of polyaluminum chloride and chitosan. When 21.0 mg/L of the coagulant is added, 98.15% turbidity, 67.78% COD, and 84.05% TP can be removed. This coagulant can be applied in the pretreatment of blue algae biogas slurry [56]. Figure 6 showed the SEM of PACl-CTS. PACl with certain crystal structure was embedded in chitosan, which enabled the composite coagulant to possess much more positive charge. When the PACl-CTS was added to the algal water, it can decrease the negative charge on the surface of microalgae and enhance the flocculation by polymer bridging.

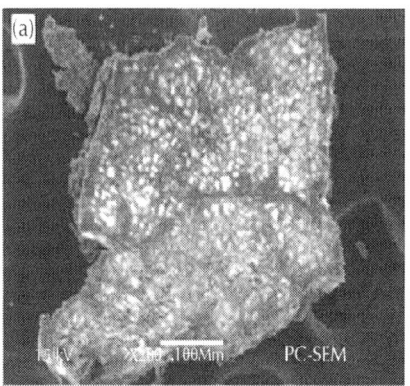

**Figure 6:** SEM of PACl-CTS [56].

## *Aluminium Sulphate Modified Chitosan*

Wang et al. used aluminium sulphate and chitosan as coagulants to treat the high algae-laden water. The compound action of aluminium sulphate and chitosan can reduce dosage of aluminum salt coagulant for meeting the treatment requirements, which in turn reduces the residual aluminium in treated water. Moreover, with the coagulation aid of chitosan, the algae flocs were larger and more compact and had a faster settling velocity [57].

## *Fly Ash Modified Chitosan*

The performance of activated fly ash modified chitosan (FA-MC) as a flocculant to remove Microcystis aeruginosa was reported by Qiao et al. [58]. 90% of algae can be removed at the dosage of 0.25 mg/L chitosan within 1 h or at the dosage of 0.35 mg/L chitosan within 40 min. The authors found that the algal extracellular organic matter had a priority to consumption of flocculant in initial period of flocculation. In the late stage of flocculation, extracellular organic matter can decrease the adsorption bridging and entrapping-weeping functions of fly ash modified chitosan.

## *Magnetic Chitosan*

Ferroferric oxide modified with chitosan was used to remove algal in freshwater [59]. This magnetic polymer could remove over 99% algal cells, which is much more effective when comparing with chitosan or $Fe_3O_4$ alone. The author found that the high algal removal efficiency of magnetic polymer is due to the cooperation between chitosan and $Fe_3O_4$ particles. Chitosan can enhance the function of netting and bridging in flocculating algal cells, while the $Fe_3O_4$ could separate the flocculated algal cells from water by its high magnetic response in the presence of magnetic field. Figure 7 showed the SEM of algal cells that was captured by magnetic polymer. The network bridge of chitosan and the magnetic $Fe_3O_4$ agglomerated the cells. After flocculation, the algae cells were still kept in original shape, which indicated that the flocculant did not destroy the cells.

**Figure 7:** SEM images of the floccules of magnetic polymer algal cells (8,000) [59].

This technique has been successfully applied to pilot experiments in Chaohu Lake, China.

# Nanochitosan

Nanoscaled chitosan particles are prepared by some researchers. Such chitosan-coated nanoparticles have overcome the limitation of microscaled particles by providing larger surface area. Owning to high adsorption capacity and stability, these particles can be used as adsorbents for food dyes adsorption [60], protein [61], and microalgae.

Farid et al. prepared chitosan nanopolymer using ionic gelation method, which added sodium tripolyphosphate to chitosan solution [62]. The average of particle size was 13.7 nm.

When nanochitosan was used as flocculant agent, the dosage of flocculant consumption decreased to 60 mg/L, while the optimum chitosan dosage was 100 mg/L for harvesting. The removal efficiency was about 98%, which increased by 9% when nanochitosan instead of chitosan was used. The author figured out that the nanochitosan had high harvesting rate because its particle had ions cross-linked with sodium tripolyphosphate. When chitosan is dissolved in water at acidic pH, it gives both hydrated amino group and − ions. The sodium tripolyphosphate (STPP, $Na_5P_3O_{10}$) presents both hydroxyl and phosphoric ions in water. These phosphoric ions of STPP can interact with − ions of chitosan, lead to crosslinking between chitosan and STPP, and result in big network of polymers, which adsorb microalgae and create a greater degree of bridging. In addition, the nanosize particles increase the adsorption ability and contact surfaces.

The author also analyzed the cost of harvesting process. For production of 1 kg of dry biomass, harvesting process would cost about $0.0246, which showed the feasibility of using nanochitosan as flocculation agent.

# CONCLUSIONS

Flocculation using chitosan is becoming a promising alternative to replace conventional flocculants in removing/harvesting microalgae. Although chitosan itself can flocculate algae, the use of some modification process may improve the properties of chitosan. Arguments on which flocculant is better in harvesting microalgae are still going on as each of the flocculants has its own advantages and disadvantages. Due to the different experimental conditions, such as pH, microalgae species, and ionic strength, it is different to compare the low-cost flocculants. Meanwhile, more studies should transfer to industrial scale.

# CONFLICT OF INTERESTS

The authors declare that there is no conflict of interests.

# ACKNOWLEDGMENTS

Financial support from the National Natural Science Foundation of China (Grant nos. 51108310 and 51036006) and support from the Project for Developing Marine Economy by Science and Technology in Tianjin (KX2010-0005) are highly appreciated.

# REFERENCES

1. C.-Y. Chen, K.-L. Yeh, R. Aisyah, D.-J. Lee, and J.-S. Chang, "Cultivation, photobioreactor design and harvesting of microalgae for biodiesel production: a critical review," Bioresource Technology, vol. 102, no. 1, pp. 71–81, 2011.

2. G. Huang, F. Chen, D. Wei, X. Zhang, and G. Chen, "Biodiesel production by microalgal biotechnology," Applied Energy, vol. 87, no. 1, pp. 38–46, 2010.

3. Y. Chisti, "Biodiesel from microalgae beats bioethanol," Trends in Biotechnology, vol. 26, no. 3, pp. 126–131, 2008.

4. I. Rawat, R. R. Kumar, T. Mutanda, et al., "Biodiesel from microalgae: a critical evaluation from laboratory to large scale production," Applied Energy, vol. 103, pp. 444–467, 2013.

5. T. M. Mata, A. A. Martins, and N. S. Caetano, "Microalgae for biodiesel production and other applications: a review," Renewable and Sustainable Energy Reviews, vol. 14, no. 1, pp. 217–232, 2010.

6. Y. Chisti, "Biodiesel from microalgae," Biotechnology Advances, vol. 25, no. 3, pp. 294–306, 2007.

7. S. A. Scott, M. P. Davey, J. S. Dennis et al., "Biodiesel from algae: challenges and prospects," Current Opinion in Biotechnology, vol. 21, no. 3, pp. 277–286, 2010.

8. H. M. Amaro, A. C. Guedes, and F. X. Malcata, "Advances and perspectives in using microalgae to produce biodiesel," Applied Energy, vol. 88, no. 10, pp. 3402–3410, 2011.

9. I. Rawat, R. R. Kumar, T. Mutanda, et al., "Biodiesel from microalgae: a critical evaluation from laboratory to large scale production," Applied Energy, vol. 103, pp. 444–467, 2013.

10. I. de Godos, H. O. Guzman, R. Soto et al., "Coagulation/ flocculation-based removal of algal-bacterial biomass from piggery wastewater treatment," Bioresource Technology, vol. 102, no. 2, pp. 923–927, 2011.

11. G. E. Molina, E.-H. Belarbi, F. G. Acién Fernández, A. Robles Medina, and Y. Chisti, "Recovery of microalgal biomass and metabolites: process options and economics," Biotechnology Advances, vol. 20, no. 7-8, pp. 491–515, 2003.

12. N. Uduman, Y. Qi, M. K. Danquah, G. M. Forde, and A. Hoadley, "Dewatering of microalgal cultures: a major bottleneck to algae-based fuels," Journal of Renewable and Sustainable Energy, vol. 2, no. 1, Article ID 012701, 2010.

13. D. Vandamme, I. Foubert, B. Meesschaert, and K. Muylaert, "Flocculation of microalgae using cationic starch," Journal of Applied Phycology, vol. 22, no. 4, pp. 525–530, 2010.

14. N.-H. Norsker, M. J. Barbosa, M. H. Vermuë, and R. H. Wijffels, "Microalgal production—a close look at the economics," Biotechnology Advances, vol. 29, no. 1, pp. 24–27, 2011.

15. E. M. Grima, E. H. Belarbia, F. G. A. Fernándeza, et al., "Recovery of microalgal biomass and metabolites: process options and economics," Biotechnology Advances, vol. 20, no. 7-8, pp. 491–515, 2003.

16. J. Morales, J. de la Noüe, and G. Picard, "Harvesting marine microalgae species by chitosan flocculation," Aquacultural Engineering, vol. 4, no. 4, pp. 257–270, 1985.

17. L. Brennan and P. Owende, "Biofuels from microalgae—a review of technologies for production, processing, and extractions of biofuels and co-products," Renewable and Sustainable Energy Reviews, vol. 14, no. 2, pp. 557–577, 2010.

18. S. Şirin, R. Trobajo, C. Ibanez, and J. Salvadó, "Harvesting the microalgae Phaeodactylum tricornutum with polyaluminum chloride, aluminium sulphate, chitosan and alkalinity-induced flocculation,"Journal of Applied Phycology, vol. 24, no. 5, pp. 1067–1080, 2011.

19. P. M. Schenk, S. R. Thomas-Hall, E. Stephens, et al., "Second generation biofuels: high-efficiency microalgae for biodiesel production," BioEnergy Research, vol. 1, pp. 20–43, 2008.

20. C. Huang, S. Chen, and J. Ruhsing Pan, "Optimal condition for modification of chitosan: a biopolymer for coagulation of colloidal particles," Water Research, vol. 34, no. 3, pp. 1057–1062, 2000.

21. T. Okuda, A. U. Baes, W. Nishijima, and M. Okada, "Improvement of extraction method of coagulation active components from Moringa oleifera seed," Water Research, vol. 33, no. 15, pp. 3373–3378, 1999.

22. M. Özacar and I. A. Şengil, "Evaluation of tannin biopolymer as a coagulant aid for coagulation of colloidal particles," Colloids and Surfaces A, vol. 229, no. 1–3, pp. 85–96, 2003.

23. A. J. Varma, S. V. Deshpande, and J. F. Kennedy, "Metal complexation by chitosan and its derivatives: a review," Carbohydrate Polymers, vol. 55, no. 1, pp. 77–93, 2004.

24. C. Y. Hu, S. L. Lo, Chang, C. L, et al., "Treatment of highly turbid water using chitosan and aluminum salts," Separation and Purification Technology, vol. 104, pp. 322–326, 2013.

25. J. M. Silva, N. Georgi, R. Costa, et al., "Nanostructured 3D constructs based on chitosan and Chondroitin Sulphate multilayers for cartilage tissue engineering," Plos ONE, vol. 8, no. 2, Article ID e55451, 2013.

26. F. Shahidi, J. K. V. Arachchi, and Y.-J. Jeon, "Food applications of chitin and chitosans," Trends in Food Science and Technology, vol. 10, no. 2, pp. 37–51, 1999.

27. K. Ganguly, T. M. Aminabhavi, and A. R. Kulkarni, "Colon targeting of 5-fluorouracil using polyethylene glycol cross-linked chitosan microspheres enteric coated with cellulose acetate phthalate,"Industrial and Engineering Chemistry Research, vol. 50, no. 21, pp. 11797–11807, 2011.

28. J. Ma and Y. Sahai, "Chitosan biopolymer for fuel cell applications," Carbohydrate Polymers, vol. 92, no. 2, pp. 955–975, 2013.

29. L. Illum, "Chitosan and its use as a pharmaceutical excipient," Pharmaceutical Research, vol. 15, no. 9, pp. 1326–1331, 1998.

30. S. A. Agnihotri, N. N. Mallikarjuna, and T. M. Aminabhavi, "Recent advances on chitosan-based micro- and nanoparticles in drug delivery," Journal of Controlled Release, vol. 100, no. 1, pp. 5–28, 2004.

31. S. C. Angadi, L. S. Manjeshwar, and T. M. Aminabhavi, "Stearic acid-coated chitosan-based interpenetrating polymer network microspheres: controlled release characteristics," Industrial and Engineering Chemistry Research, vol. 50, no. 8, pp. 4504–4514, 2011.

32. P. Miretzky and A. F. Cirelli, "Fluoride removal from water by chitosan derivatives and composites: a review," Journal of Fluorine Chemistry, vol. 132, no. 4, pp. 231–240, 2011.

33. T. Yui, K. Imada, K. Okuyama, Y. Obata, K. Suzuki, and K. Ogawa, "Molecular and crystal structure of the anhydrous form of chitosan," Macromolecules, vol. 27, no. 26, pp. 7601–7605, 1994.

34. M. Rinaudo, "Characterization and properties of some polysaccharides used as biomaterials,"Macromolecular Symposia, vol. 245-246, no. 1, pp. 549–557, 2006.

35. D. M. Ruthven, Encyclopedia of Separation Technology, vol. 1 of A Kirk-Othmer Encyclopedia, John Wiley & Sons, New York, NY, USA, 1997.

36. X. Wu, X. Ge, D. Wang, and H. Tang, "Distinct coagulation mechanism and model between alum and high Al13-PACl," Colloids and Surfaces A, vol. 305, no. 1-3, pp. 89–96, 2007.

37. F. Renault, B. Sancey, P.-M. Badot, and G. Crini, "Chitosan for coagulation/flocculation processes—an eco-friendly approach," European Polymer Journal, vol. 45, no. 5, pp. 1337–1348, 2009.

38. N. Rashid, M. S. Rehman, and J. I. Han, "Use of chitosan acid solutions to improve separation efficiency for harvesting of the microalga Chlorella vulgaris," Chemical Engineering Journal, vol. 226, pp. 238–242, 2013

39. E. S. Beach, M. J. Eckelman, Z. Cui, et al., "Preferential technological and life cycle environmental performance of chitosan flocculation for harvesting of the green algae Neochloris oleoabundans,"Bioresource Technology, vol. 121, pp. 445–449, 2012.

40. R. Divakaran and V. N. Sivasankara Pillai, "Flocculation of river silt using chitosan," Water Research, vol. 36, no. 9, pp. 2414–2418, 2002.

41.  A. L. Ahmad, N. H. Mat Yasin, C. J. C. Derek, and J. K. Lim, "Optimization of microalgae coagulation process using chitosan," Chemical Engineering Journal, vol. 173, no. 3, pp. 879–882, 2011.

42.  Y.-S. Cheng, Y. Zheng, J. M. Labavitch, and J. S. Vandergheynst, "The impact of cell wall carbohydrate composition on the chitosan flocculation of Chlorella," Process Biochemistry, vol. 46, no. 10, pp. 1927–1933, 2011.

43.  Y. N. Xu, S. Purton, and F. Baganz, "Chitosan flocculation to aid the harvesting of the microalga Chlorella sorokiniana," Bioresource Technology, vol. 129, pp. 296–301, 2013.

44.  R. Henderson, S. A. Parsons, and B. Jefferson, "The impact of algal properties and pre-oxidation on solid-liquid separation of algae," Water Research, vol. 42, no. 8-9, pp. 1827–1845, 2008.

45.  Y. R. Chang and D. J. Lee, "Coagulation-membrane filtration of Chlorella vulgaris at different growth phases," Drying Technology, vol. 30, no. 11-12, pp. 1317–1322, 2012.

46.  M. Heasman, J. Diemar, W. O›Connor, T. Sushames, and L. Foulkes, "Development of extended shelf-life microalgae concentrate diets harvested by centrifugation for bivalve molluscs—a summary,"Aquaculture Research, vol. 31, no. 8-9, pp. 637–659, 2000.

47.  G. G. D›Ayala, M. Malinconico, and P. Laurienzo, "Marine derived polysaccharides for biomedical applications: chemical modification approaches," Molecules, vol. 13, no. 9, pp. 2069–2106, 2008.

48.  G. Pan, M.-M. Zhang, H. Chen, H. Zou, and H. Yan, "Removal of cyanobacterial blooms in Taihu Lake using local soils. I. Equilibrium and kinetic screening on the flocculation of Microcystis aeruginosa using commercially available clays and minerals," Environmental Pollution, vol. 141, no. 2, pp. 195–200, 2006

49.  G. Pan, H. Zou, H. Chen, and X. Yuan, "Removal of harmful cyanobacterial blooms in Taihu Lake using local soils. III. Factors affecting the removal efficiency and an in situ field experiment using chitosan-modified local soils," Environmental Pollution, vol. 141, no. 2, pp. 206–212, 2006.

50.    G. Pan, J. Chen, and D. M. Anderson, "Modified local sands for the mitigation of harmful algal blooms,"Harmful Algae, vol. 10, no. 4, pp. 381–387, 2011.

51.    G. Pan, L. Dai, L. Li et al., "Reducing the recruitment of sedimented algae and nutrient release into the overlying water using modified soil/sand flocculation-capping in eutrophic lakes," Environmental Science and Technology, vol. 46, no. 9, pp. 5077–5084, 2012.

52.    H. Zou, G. Pan, H. Chen, and X. Yuan, "Removal of cyanobacterial blooms in Taihu Lake using local soils. II. Effective removal of Microcystis aeruginosa using local soils and sediments modified by chitosan," Environmental Pollution, vol. 141, no. 2, pp. 201–205, 2006.

53.    X. Yang, C. Wu, Y. He, B. Zhang, and F. Li, "Removal effects and mechanisms of Microcystis aeruginosaby Chitosan-modified adsorbent," in Proceedings of the 2nd International Symposium on Aqua Science, Water Resource and Low Carbon Energy, vol. 1251 of AIP Conference Proceedings, pp. 125–128, December 2009.

54.    J. Shao, Z. Wang, Y. Liu et al., "Physiological responses of Microcystis aeruginosa NIES-843 (cyanobacterium) under the stress of chitosan modified kaolinite (CMK) loading," Ecotoxicology, vol. 21, no. 3, pp. 698–704, 2011.

55.    Y. H. Wang, S. G. Zhou, and Y. X. Yang, "Application of chitosan in removing algae by dissolved air flotation," in Advanced Materials Research, G. Li, Y. Huang, and C. Chen, Eds., vol. 347–353, pp. 1911–1916, 2011.

56.    W. Zhang, P. Fan, Q. Li, et al., "Synthsis of PACl-CTS composite coagulant and application in the pre-treatment of the blue algae biogas slurry," Environmental Chemistry, vol. 31, no. 7, pp. 1057–1062, 2012.

57.    Y. H. Wang, S. G. Zhuo, X. Y. Zhou, et al., "Improvement of high algae-laden water treatment by coagulation aid of chitosan," in Advanced Materials Research, G. Li, Y. Huang, and C. Chen, Eds., vol. 250–253, pp. 3454–3459, 2011.

58.    J. Qiao, L. Dong, and Y. Hu, "Removal of harmful algal blooms using activated fly ash-modified chitosan," Fresenius Environmental Bulletin A, vol. 20, no. 3, pp. 764–772, 2011.

59. D. Liu, F. Li, and B. Zhang, "Removal of algal blooms in freshwater using magnetic polymer," Water Science and Technology, vol. 59, no. 6, pp. 1085–1091, 2009.

60. Z. K. Zhou, S. Q. Lin, T. L. Yue, et al., "Adsorption of food dyes from aqueous solution by glutaraldehyde cross-linked magnetic chitosan nanoparticles," Journal of Food Engineering, vol. 126, pp. 133–141, 2014.

61. J. S. Gonzalez, P. Nicolás, M. L. Ferreira, et al., "Fabrication of ferrogels using different magnetic nanoparticles and their performance on protein adsorption," Polymer International, vol. 63, no. 2, pp. 258–265, 2014.

62. M. S. Farid, A. Shariati, A. Badakhshan, et al., "Using nano-chitosan for harvesting microalgaNannochloropsis sp," Bioresource Technology, vol. 131, pp. 555–559, 2013.

63. R. Divakaran and V. N. S. Pillai, "Flocculation of algae using chitosan," Journal of Applied Phycology, vol. 14, no. 5, pp. 419–422, 2002.

64. B. Riaño, B. Molinuevo, and M. C. García-González, "Optimization of chitosan flocculation for microalgal-bacteria biomass harvesting via response surface methodology," Ecological Engineering, vol. 38, no. 1, pp. 110–113, 2012.

65. Y. M. Chen, J. C. Liu, and Y.-H. Ju, "Flotation removal of algae from water," Colloids and Surfaces B: Biointerfaces, vol. 12, no. 1, pp. 49–55, 1998.

# A Study on Ethanolysis and Methanolysis of Coconut Oil for Enzymatically Catalyzed Production of Biodiesel

Livia M. O. Ribeiro, Albanise E. Silva, Margarete C. S. Silva, Renata M. R. G. Almeida*

Tecnology Center, Federal University of Alagoas, Avenida Lourival Melo Mota, Tabuleiro do Martins, Maceió, Brazil

## ABSTRACT

Biodiesel production by enzymatic catalysis has been the subject of much research for developing processes that can potentially compete with other types of catalysis. The objective of this study was to investigate the variables that influenced the transesterification of coconut oil catalyzed by immobilized lipase for biodiesel production. A full $2^4$ factorial design with the variables temperature (40°C - 60°C), enzyme concentration (3% - 7%), oil-ethanol ratio (1:6 - 1:10) and alcohol type (methanol-ethanol) was performed. The best conversion result (80.5%) was obtained using ethanol with a higher temperature, molar ratio and enzyme concentration. The obtained yields showed that the results attained with ethanol were more significant when compared with methanol.

# INTRODUCTION

The demand for renewable fuels has recently been much intensified. In this scenario, biodiesel arises as an alternative to petroleum products, aiming to reduce pollutant emissions in the atmosphere.

Biodiesel is the name of a clean-burning fuel produced from domestic renewable resources. It does not contain petroleum, but can be added to it forming a blend [1]. It can be used in a compression-ignition engine (diesel) without need for modification. Biodiesel is simple to use, biodegradable, non-toxic and essentially free of sulfur and aromatic compounds [2].

To produce biodiesel, vegetable oils are mixed with short-chain alcohols (ethanol or methanol) and then stimulated by a catalyst. This chemical process is known as transesterification, where glycerin is separated from the fat or oil. The process generates two products: esters (the chemical name for biodiesel) and glycerin (a valuable product in the soap industry). The oil is separated from the glycerin by filtration process [3].

High-quality biodiesel must be produced following strict industry specifications, as the ASTM D6751 at international level. In USA, biodiesel is the only alternative fuel to obtain full approval in the Clean Air Act of 1990, which is authorized for sale and distribution by the USA Environmental Agency (EPA) [4].

A very appropriate nomenclature has been used internationally to identify the concentration of biodiesel in the blend, defined as BX, wherein X refers to the percentage by volume of biodiesel. Therefore, B2, B5 and B20 refer to fuels at concentrations of 2%, 5% and 20% of added biodiesel, respectively [5]

Although soybean oil is the most widely used in biodiesel production, other oilseed plants with a higher oil content, such as peanut, sunflower, corn, canola (rapeseed), castor oil and cotton have been reported in the literature as favorable for the production of biodiesel [6] - [8] . In this context, coconut oil has been identified as an important potential source of raw material for the local production of biodiesel in the northeast of Brazil [9].

In the transesterification of vegetable oil, a triglyceride reacts with an alcohol in the presence of a catalyst, which may be an acid, a strong base or an enzyme, producing a mixture of alkyl esters of fatty acids and glycerol. The total process is a sequence of three consecutive and reversible reactions, where diglycerides and monoglycerides are formed as intermediates. For a complete stoichiometric transesterification, a 3:1 molar ratio of alcohol to triglyceride is required. Due to the reaction's reversible character, a transesterification agent (alcohol) is usually added in excess, thus contributing to increasing the ester yield, and to enabling its separation from the glycerol formed [10].

Currently, the commercial production of biodiesel occurs via a chemical route, but the enzymatic process has attracted the interest of the scientific community. Although the biodiesel production through enzymatic transesterification has not yet been commercially developed, advances in this process have been reported in papers and patents. The common aspect of these studies is the optimization of reaction conditions (solvent, temperature, pH, type of microorganism that produces the enzyme, immobilization technologies, etc.), in order to establish the characteristics for industrial applications [11].

However, once the enzymatic process is optimized, it may present some advantages over the chemical process (Table 1).

Methanol is used as the acyl group receptor on biodiesel production by transesterification of oils or fats using lipases, according to most recent studies found in the literature. Methanol is more reactive, and produces more volatile esters and it is cheaper on the international market, when compared with other short chain alcohols [12].

However, some authors are considering other alcohols, such as ethanol, propanol, isopropanol and butanol for the production of biodiesel component esters, due to methanol's toxicity and by the fact that it is produced from petroleum. Ethanol is preferred because it is considered renewable [13].

The objective of this work was to study the variables that influenced the transesterification of coconut oil in the production of biodiesel catalyzed by immobilized lipase. The transesterification was performed in sealed glass reactors stirred at 200 rpm and catalyzed by the commercial immobilized lipase Novozym 435 using ethanol and methanol. Assays were carried out in accordance with an experimental design where the variables were temperature ($40°C$ - $60°C$), enzyme concentration (3% - 7%), ratio oil-alcohol (1:6 - 1:10) and type of alcohol (ethanol and methanol).

**Table 1:** .Advantages and disadvantages of the chemical and enzymatic processes in biodiesel production

| Process | Advantages | Disadvantages |
|---------|-----------|---------------|
| Chemical | Simplicity; High yield; Short reaction time. | Difficulty of separating the catalyst; Impossibility of reusing the catalyst; Difficulty of use hydrated ethanol; Obtaining products with lower purity. |
| Enzymatic | Ease of catalyst (support) separation; Obtaining products with higher purity; Possibility of using hydrated ethanol in the reaction. | Long reaction time; Enzymes cost; Low yield. |

# MATERIALS AND METHODS

## Materials

The samples of crudecoconutoil were provided by SOCOCO Coconut Food Industry, located in Maceió, Brazil. According to the supplier, the extraction of coconut oil was made by mechanical pressing of the dehydrated pulp.

Enzymes used in this study were provided by Novozymes, Latin America (Brazil). Other reagents used were of analytical grade.

# Characterization of the Transesterification Products

To determine the biodiesel yield (%), the characterization of methyl or ethyl esters was carried out using gas chromatography (GC). Transesterification products were analyzed by gas chromatography using a VARIANCP- 3800 instrument equipped with a FID detector (Flame Ionization Detection) and a short capillary column (2.3 m). The temperature of the detector was 250°C and the injector was 240°C. The oven had temperature programmed from 150 to 260°C at a heating rate (ramp) of 10°C/min. Tricaprylin was used as the internal standard and hydrogen of high purity (99.95%) was used as the carrier gas.

The analyzed sample was prepared by mixing 0.15 mL of biodiesel previously purified with 1 mL of standard solution (tricaprylin plus hexane in desiccator). A 1 µL aliquot of the sample was then injected into the chromatograph, with a 10 mL glass syringe.

The yield calculation in esters was carried out based on the masses and areas under the peaks corresponding to the methyl or ethyl esters and to the internal standard, using Equation (1):

$$Yield(\%) = \frac{m_p A_b f}{m_b A_p} \times 100 \qquad (1)$$

Where:

$m_p$ = internal standard weight (0.08 g);

$A_b$ = sum of the peak are as referring to the esters in the sample (peak detected between 8 and 13 min);

f = response factor (0.78);

$A_p$ = peak area referring to the internal standard (tricaprylin plus hexane- peak detected between 15 and 18 min);

$m_b$ = sample weight (0.15 g).

The conversion analyses were performed in duplicate. The average conversion for each experiment was also calculated.

# EXPERIMENTAL STAGE

## Enzymatic Transesterification

The reactions, in laboratory scale, were carried out in 250 mL Erlenmeyer flasks closed with glass lids. The commercial lipase in pre-established concentrations was added to the oil-alcohol mixture. The flasks were then incubated in a rotating chamber at 200 rpm at a controlled temperature. The amount of coconut oil used was held at 25 g.

After the elapsed reaction time, the sample was filtered (in order to retain the enzymes for subsequent recovery and use) and then taken to a rotary evaporator at 90°C at 40 rpm, to evaporate the alcohol.

After this procedure, the resulting reaction mixture was placed in a separatory funnel, where the upper phase, concentrated in methyl or ethyl esters (biodiesel), was separated from the lower phase containing the remaining reaction products (mono and diglycerides, glycerol and impurities), as well as the unreacted substrates (alcohol and oil).

## Experimental Design

The experimental design is a very useful tool for laboratory research. In addition to allowing research with several variables, it also enables an easy determination of the variation effects, since determining the influence of one or more variables over the others is a common problem.

A full factorial experimental design with two levels and four variables was performed to determine the experimental conditions that maximize the synthesis of biodiesel resulting from the reaction and to evaluate the influence of selected variables.

The variables studied at this stage were: oil-alcohol molar ratio, temperature, and enzyme percentage and alcohol type. Based on preliminary studies, the agitation was kept constant, since it has been verified that no change results when increasing it from 200 to 500 rpm [14]. The reaction time was set at 24 hours. The limitations associated with each variable are shown in Table 2.

The experimental matrix for the $2^4$ factorial design is presented in Table 3.

Through the obtained experimental data, an empirical model was developed with the variables coded by the letter x. This model was used to estimate the yield of coconut oil biodiesel by using coded values of the variables here studied. These coded values must belong to the interval $-1 \leq x \leq +1$, which was considered in this work.

Therefore, the following mathematical model, Equation (2), for the prediction of desired responses was adopted:

$$y = \beta_0 + \sum_{j=1}^{x} \beta_j x_j + \sum_{i<j}^{x} \beta_{ij} x_j x_i \qquad (2)$$

Where:

y = percentage of the variable of interest (yield);

x = variable that shows the factors in coded scales. As in the present work only the low and high levels are being considered, x assumes $-1$ and $+1$ values;

= parameters which values will be determined.

For practical purposes, the $_0$ estimate is the general average of the responses, while the estimates of the other parameters were obtained from the regression table given by the STATISTICA7.0.

**Table 2:**-Range of variables

| Variables | Extreme Negative (−1) | Extreme Positive (+1) |
|---|---|---|
| Molar ratio—oil:alcohol [R] | 1:6 | 1:10 |
| Temperature [T] (˚C) | 40 | 60 |
| Enzyme percentage [E] (%p/p) | 3 | 7 |
| Alcohol type | Ethanol | Methanol |

**Table 3**: Experimental design matrix

| Assay | Type | R (O:Alcohol) | T (°C) | E (%p/p) |
|---|---|---|---|---|
| 1 | Ethanol/−1 | 1:6/−1 | 40/−1 | 3/−1 |
| 2 | Ethanol/−1 | 1:6/−1 | 40/−1 | 7/+1 |
| 3 | Ethanol/−1 | 1:6/−1 | 60/+1 | 3/−1 |
| 4 | Ethanol/−1 | 1:6/−1 | 60/+1 | 7/+1 |
| 5 | Ethanol/−1 | 1:10/+1 | 40/−1 | 3/−1 |
| 6 | Ethanol/−1 | 1:10/+1 | 40/−1 | 7/+1 |
| 7 | Ethanol/−1 | 1:10/+1 | 60/+1 | 3/−1 |
| 8 | Ethanol/−1 | 1:10/+1 | 60/+1 | 7/+1 |
| 9 | Methanol/+1 | 1:6/−1 | 40/−1 | 3/−1 |
| 10 | Methanol/+1 | 1:6/−1 | 40/−1 | 7/+1 |
| 11 | Methanol/+1 | 1:6/−1 | 60/+1 | 3/−1 |
| 12 | Methanol/+1 | 1:6/−1 | 60/+1 | 7/+1 |
| 13 | Methanol/+1 | 1:10/+1 | 40/−1 | 3/−1 |
| 14 | Methanol/+1 | 1:10/+1 | 40/−1 | 7/+1 |
| 15 | Methanol/+1 | 1:10/+1 | 60/+1 | 3/−1 |
| 16 | Methanol/+1 | 1:10/+1 | 60/+1 | 7/+1 |

# RESULTS AND DISCUSSION

## Transesterification Reactions

The experimental data obtained in the enzymatic transesterification of coconut oil using the commercial lipase Novozym 435 as a catalyst are shown in Table 4. Yield data referred to a 24-hour reaction. It is noteworthy that the experiments were done induplicate at each point.

Table 4 shows that the highest yield (80.5%) was obtained at the higher temperature (60°C), enzyme concentration (7%), molar ratio of oil:alcohol (1:10) and using ethanol.

Once the yield was determine, a statistical model was built for representing the experimental data, assessing the significance of the variables, and evaluating possible interactions between them.

The statistical model for calculating the yield of coconut biodiesel can be represented by Equation (3):

$$y = 54.97 + 2.57x_1 - 11.24x_2 + 4.35x_3 - 14.64x_4 - 3.82x_2x_3 - 15.14x_2x_4 - 2.66x_3x_4$$
(3)

Where:

The regression parameters 2.57, −11.24, 4.35, −14.64, −3.82, −15.14, and −2.66 were obtained from Table 5, and the parameter 54.97 is the mean of all responses obtained from the experiments:

$y$ is the yield (response) estimate percentage;

$x_1$ is the coded variable of enzyme concentration which belongs to the interval $-1 \leq x \leq 1$;

$x_2$ is the coded variable of temperature which belongs to the interval $-1 \leq x \leq 1$;

$x_3$ is the coded variable of molar ratio which belongs to the interval $-1 \leq x \leq 1$;

$x_4$ is the coded variable of alcohol type which belongs to the interval $-1 \leq x \leq 1$.

The analysis of variance (ANOVA) (Table 6) shows that the model used in this study is statistically significant and suitable to represent the relation between response and the significant variables, because the coefficient of determination is satisfactory ($R^2 = 0.98$). This value indicates that 98% of the variation of the yield can be attributed to the independent variables and only 2% of the total variation cannot be explained by the model, suggesting that there was a good fit of the model to the experimental data.

Linearity and good fit of the model are shown in Figure 1 with a confidence interval of 95%.

To a significance level of 5%, the obtained results indicate that enzyme concentration, temperature, molar ratio, type of alcohol, as well as interactions between temperature and molar ratio, temperature and type of alcohol, and molar ratio and type of alcohol are significant. These significant values ($p < 0.05$) are best viewed through

**Table 4**: Yields obtained in the transesterification of coconut oil

| Assay | Type | R (O:Alcohol) | T (°C) | E (%p/p) | 1st Run | 2nd Run | Yield (%) |
|---|---|---|---|---|---|---|---|
| 1 | Ethanol | 1:6 | 40 | 3 | 48 | 50.2 | 49.1 |
| 2 | Ethanol | 1:6 | 40 | 7 | 61 | 62.8 | 61.9 |
| 3 | Ethanol | 1:6 | 60 | 3 | 61.5 | 65.1 | 63.3 |
| 4 | Ethanol | 1:6 | 60 | 7 | 76.9 | 75.3 | 76.1 |
| 5 | Ethanol | 1:10 | 40 | 3 | 79.3 | 76.7 | 78 |
| 6 | Ethanol | 1:10 | 40 | 7 | 72.2 | 75.5 | 73.8 |
| 7 | Ethanol | 1:10 | 60 | 3 | 77.6 | 70.6 | 74.1 |
| 8 | Ethanol | 1:10 | 60 | 7 | 84 | 77.1 | 80.5 |
| 9 | Methanol | 1:6 | 40 | 3 | 58.4 | 55.9 | 57.1 |
| 10 | Methanol | 1:6 | 40 | 7 | 68.4 | 59.7 | 64 |
| 11 | Methanol | 1:6 | 60 | 3 | 19.2 | 19.6 | 19.4 |
| 12 | Methanol | 1:6 | 60 | 7 | 14.2 | 13.8 | 14 |
| 13 | Methanol | 1:10 | 40 | 3 | 68.2 | 67.2 | 67.7 |
| 14 | Methanol | 1:10 | 40 | 7 | 77.9 | 78.1 | 78 |
| 15 | Methanol | 1:10 | 60 | 3 | 10.3 | 10.6 | 10.4 |
| 16 | Methanol | 1:10 | 60 | 7 | 12.5 | 11.4 | 12 |

**Table 5**. Regression coefficients for the biodiesel yield

| Factors | Regression Coefficient | Standard Error | T | p | Confidence Interval −95% | Confidence Interval +95% |
|---|---|---|---|---|---|---|
| Mean | 54.97 | 0.79 | 69.42 | 0.0000 | 53.32 | 56.62 |
| (1) E% | 2.57 | 0.79 | 3.25 | 0.004 | 0.93 | 4.22 |
| (2) Temperature | −11.24 | 0.79 | −14.2 | 0.0000 | −12.89 | −9.6 |
| (3) R molar | 4.35 | 0.79 | 5.49 | 0.0000 | 2.7 | 5.99 |
| (4) Type | −14.64 | 0.79 | −18.48 | 0.0000 | −16.28 | −12.99 |
| 1 × 2 | −0.65 | 0.79 | −0.83 | 0.4166 | −2.3 | 0.99 |
| 1 × 3 | −0.81 | 0.79 | −1.07 | 0.3165 | −2.46 | 0.83 |
| 1 × 4 | −0.91 | 0.79 | −1.15 | 0.2621 | −2.56 | 0.73 |
| 2 × 3 | −3.82 | 0.79 | −4.82 | 0.0000 | −5.46 | −2.17 |
| 2 × 4 | −15.14 | 0.79 | −19.12 | 0.0000 | −16.79 | −13.5 |
| 3 × 4 | −2.66 | 0.79 | −3.36 | 0.0029 | −4.31 | −1.02 |

**Table 6**. Analyses of variance (ANOVA)

| Variation Source | Sum of Squares | Mean Square | F | p |
|---|---|---|---|---|
| (1) E% | 212.18 | 212.18 | 10.57 | 0.0038 |
| (2) Temperature | 4045.50 | 4045.50 | 201.62 | 0.0000 |
| (3) R molar | 605.52 | 605.52 | 30.18 | 0.0000 |
| (4) Type | 6856.21 | 6856.21 | 341.7 | 0.0000 |
| 2 by 3 | 466.65 | 466.65 | 23.26 | 0.0000 |
| 2 by 4 | 7338.66 | 7338.66 | 365.74 | 0.0000 |
| 3 by 4 | 226.85 | 226.85 | 11.31 | 0.0000 |
| Error | 421.36 | 20.065 | | |
| Total | 20234.48 | | | |

$R^2 = 0.98$

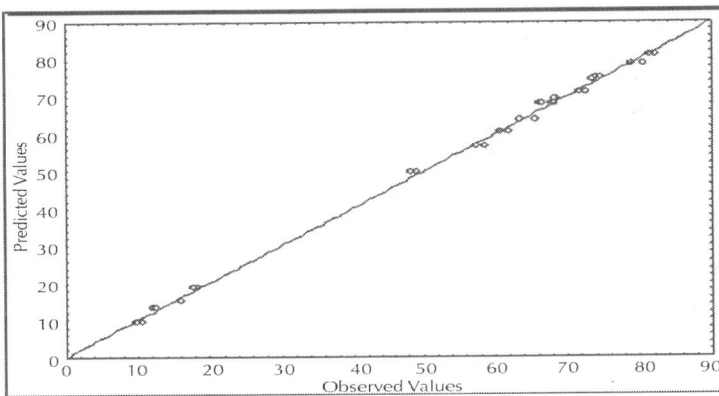

**Figure 1**: Confidence interval between the model and the experimental data at 95%.

The Pareto chart in Figure 2:

The values present in the Pareto chart represent the individual influence of each variable on the system response (the yield, in this case) and are interpreted as follows: The reaction yield decreases 30.29% on average when the temperature changes from the lower (40°C) to the higher level (60°C) and when ethanol (lower level) is exchanged for methanol (higher level);

The reaction yield decreases 29.27% on average when using methanol (higher level) instead of ethanol (lower level);

The yield of the reaction decreases 22.49% on average when the temperature changes from the lower (40°C) to the upper level (60°C);

The yield of the reaction increases 8.7% on average when the molar ratio changes from the lower (1:6) to the upper level (1:10);

The reaction yield decreases 7.64% on average when the temperature changes from the lower (40°C) to the top level (60°C) and the molar ratio changes from the lower (1:6) to the top level (1:10);

The reaction yield decreases 5.32% on average when the molar ratio changes from the lower (1:6) to the upper level (1:10) and when ethanol (lower level) is replaced by methanol (top level);

The reaction yield increases 5.15% on average, when the enzyme concentration changes from the lower (3%) to the upper level (7%).

The interactions between enzyme concentration and type of alcohol, enzyme concentration and molar ratio, and enzyme concentration and temperature had no significant effects, since $p > 0.05$. Although not significant, this result provides very interesting and attractive information. One of the problems involving enzymatic catalysis is the enzyme costs. A smaller amount of enzyme would make the process more feasible and appealing. The results obtained in this study show that the yield is not significantly influenced when 3% or 7% of enzyme are used. Hence, coconut oil biodiesel can be produced using enzyme concentrations below 7%, and still provide good yields.

# Influence of Reaction Parameters

The higher the enzyme concentration, the greater the reaction yield [15]. In this study, the enzyme concentration had a positive effect, showing a higher yield when a greater amount of enzyme was used. However, when the assays 7 and 8 are compared it is notice that using 3% of enzyme, the yield did not decrease significantly; this variable can be optimized in future studies, keeping the other variables in the best conditions.

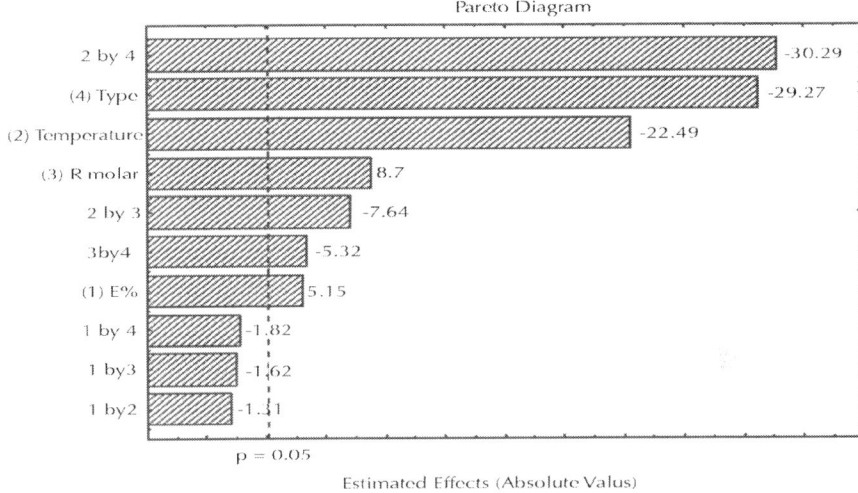

Figure 2: Pareto diagram with a significance level of 5%.

The molar ratio of substrate is a variable with great influence on the biodiesel synthesis reaction. Excess alcohol in the 3:1 stoichiometric ratio is used to ensure a high reaction rate and minimize diffusion limitations. However, excessive levels of alcohol can inhibit the enzyme activity, thus decreasing its catalytic activity during the transesterification reaction. Therefore, a high oil-alcohol ratio means a higher medium polarity produced by alcohol and water. This inactivates the enzyme and consequently the alkyl esters yield [16].

As two types of alcohol were used in this study, these influences were evaluated separately.

When using ethanol, the molar ratio had a positive effect, demonstrating no enzyme inhibition for alcohol excess. This result corroborates [17] who observed similar effects when using castor oil.

Regarding the methanol, the molar ratio had negative effect, showing that the smaller molar ratio (1:6) yielded better results. This can be explained by the fact that there is a greater deactivation of enzymes by this alcohol [18]

The reaction rate increases with temperature. In this work, at a temperature range between 40°C and 60°C, enzyme activity increased with temperature. However, when the temperature interacts with an alcohol, this behavior cannot be generalized. Temperature affects differently each type of alcohol. When using ethanol, the temperature had a positive effect, showing no enzyme inhibition at high temperatures.

When ethanol is replaced by highly volatile methanol, a significant yield drop was noticed. This can most likely occur due to the boiling point of methanol (64.70°C) to be very close to the maximum temperature used (60°C).

# Influence of Alcohol Type

Methanol used in the transesterification reaction did not produce good yields when compared to ethanol. The results obtained for this alcohol corroborate those found in the literature. Methanol is a highly hydrophilic solvent, thus being able to solubilize and remove the essential water layer surrounding the enzymes, which can lead to a loss of catalytic activity of the lipase [16] - [19] . According to [19] the poor results obtained with methanol may be related to the fact that the material used to immobilize enzymes (acrylic resin) can absorb polar compounds such as methanol. When the alcohol concentration in the mixture is high, it forms droplets that bind with the resin particles. The binding of alcohol molecules to the enzymes blocks the reaction (input) with triglycerides, resulting in low yields.

The use of ethanol in transesterification reactions to obtain coconut biodiesel had a positive effect. This result is very interesting and attractive, because ethanol has some advantages in relation to methanol, including: lower toxicity, renewability and availability in the region of this study.

Once discovered the best conditions for the production of biodiesel within the range studied in this work, it was possible to obtain the response surface for the yield using molar ratio and temperature as independent variables (Figure 3). The other two variables (enzyme concentration and type of alcohol) were fixed at their best results (7% and ethanol). Figure 3 shows that by using ethanol and 7% of enzyme, the yield increases when a molar ratio of 1:10 (+1) at 60°C (+1) is used.

The best yield obtained in this study was 80.5% with 60°C, using 7% of the enzyme, molar ratio oil-alcohol 1:10 and ethanol and this yield is too low when compared with alkali-catalysed process and the reaction rates are appreciably slower. Although, enzyme-based process has other advantages such as the capable of minimizing saponification problems due to a relatively high free fatty acid content [20] . Others papers that investigated biodiesel from coconut oil enzymatically catalyzed has obtained yield values between 35 and 90% [20] - [22]

# CONCLUSIONS

Biodiesel production from coconut oil was conducted by enzymatic catalysis in the presence of methanol or ethanol. The process conditions (temperature, molar ratio oil-alcohol, enzymeconcentration and type of alcohol) were varied in order to check their influence. To this end, a full $2^4$ factorial design was performed.

This experimental design has proven efficient to study the influence of the process variables. The effects of temperature, type of alcohol, molar ratio, enzyme concentration and interactions between temperature and molar ratio, temperature and type of alcohol and the molar ratio and type of alcohol were very significant. The model was adjusted to the factorial design responses, showing the characteristics of a significant model under the statistical point of view, and the model has also predictive characteristics within the range studied for each variable.

The best yield in this study (80.5%) was obtained with 60°C, using 7% of the enzyme, molar ratio oil: alcohol of 1:10 and ethanol.

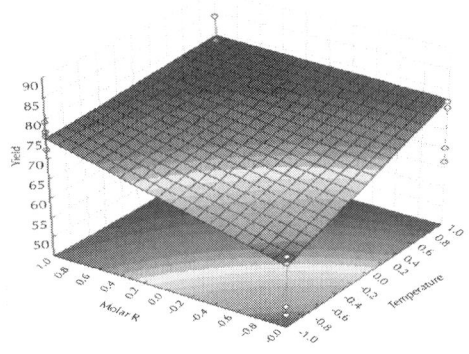

**Figure 3:** Response surface to yield in biodiesel as a function of the molar ratio and temperature.

The results obtained when using methanol were not as significant when compared with ethanol. The results obtained using ethanol were very attractive, because ethanol was renewable and easily obtained in the region of this study.

# REFERENCES

1. Zhang, Y., Dubé, M.A., Mclean, D.D. and Kates, M. (2003) Biodiesel Production from Waste Cooking Oil: Process Design and Technological Assessment. Bioresource Technology, 89, 1-16. http://dx.doi.org/10.1016/S0960-8524(03)00040-3

2. Worapun, I., Pianthong, K. and Thaiyasuit, P. (2012) Optimization of Biodiesel Production from Crude Palm Oil Using Ultrasonic Irradiation Assistance and Response Surface Methodology. Journal of Chemical Technology and Biotechnology, 87, 189-197.http://dx.doi.org/10.1002/jctb.2679

3. Kumar, D., Kumar, G. and Singh, P.C.P. (2010) Fast, Easy Ethanolysis of Coconut Oil for Biodiesel Production Assistedby Ultrasonication. Ultrasonics Sonochemistry, 17, 555-559.http://dx.doi.org/10.1016/j.ultsonch.2009.10.018

4. Ma, F. and Hanna, M.A. (1999) Biodiesel Production: A Review. Bioresource Technology, 70, 1-15. http://dx.doi.org/10.1016/S0960-8524(99)00025-5

5. Fukuda, H., Kondo, A. and Noda, H. (2001) Biodiesel Fuel Production by Transesterification of Oils. Journal of Bioscience and Bioengineering, 92, 405-416.http://dx.doi.org/10.1016/S1389-1723(01)80288-7

6. dos Santos, I.C.F., de Carvalho, S.H.V., Solleti, J.I., de La Salles, W.F., La Salles, K.T.S. and Meneghetti, S.M.P. (2008) Studies of Terminalia catappa L. Oil: Characterization and Biodiesel Production. Bioresource Technology, 99, 6545- 6549.http://dx.doi.org/10.1016/j.biortech.2007.11.048

7. Silva, J.P.V., Serra, T.M., Gossmann, M., Wolf, C.R., Meneghetti, M.R. and Meneghetti, S.M.P. (2010) Moringa Oleifera Oil: Studies of Characterization and Biodiesel Production. Biomass Bioenergy, 34, 1527-1530. http://dx.doi.org/10.1016/j.biombioe.2010.04.002

8. La Salles, K.T.S., Meneghetti, S.M.P., La Salles, W.F., Meneghetti, M.R., dos Santos, I.C.F., da Silva, J.P.V., de Carvalho, S.H.V. and Soletti, J.I. (2010) Characterization of Syagrus coronata (Mart.) Becc. Oil and Properties of Methyl Esters for Use as Biodiesel. Industrial Crops and Products, 32, 518-521.http://dx.doi.org/10.1016/j.indcrop.2010.06.026

9. Ribeiro, L.M.O., Santos, B.C.S. and Almeida, R.M.R.G. (2012) Studies on Reaction Parameters Influence on Ethanolic Production of Coconut Oil Biodiesel Using Immobilized Lipase as a Catalyst. Biomass and Bioenergy, 47, 498-503.http://dx.doi.org/10.1016/j.biombioe.2012.09.041

10. Murugesan, A., Umarani, C., Chinnusamy, T.R., Krishnan, M., Subramanian, R. and Neduzchezhain, N. (2009) Production and Analysis of Bio-Diesel from Non-Edible Oils—A Review. Renewable and Sustainable Energy Reviews, 13, 825- 834.http://dx.doi.org/10.1016/j.rser.2008.02.003

11. Tan, T., Lu, J., Nie, K., Deng, L. and Wang, F. (2010) Biodiesel Production with Immobilized Lipase: A Review. Biotechnology Advances, 28, 628-634.http://dx.doi.org/10.1016/j.biotechadv.2010.05.012

12. Talukder, M.M.R., Das, P., Fang, T.S. and Wu, J.C. (2011) Enhanced Enzymatic Transesterification of Palm Oil to Biodiesel. Biochemical Engineering Journal, 55, 119-122.http://dx.doi.org/10.1016/j.bej.2011.03.013

13. Akoh, C.C., Chang, S., Lee, G. and Shaw, J. (2007) Enzymatic Approach to Biodiesel Production. Journal of Agricultural and Food Chemistry, 55, 8995-9005.http://dx.doi.org/10.1021/jf071724y

14. Oliveira, D. and Alves, T.L.M. (2000) A Kinetic Study of Lipase Catalyzed Alcoholysis of Palm Kernel Oil. Applied Biochemistry and Biotechnology, 84-86, 59-68.

15. Oliveira, D., Oliveira, J.V., Faccio, C., Menoncin, S. and Amroginski, C. (2004) Influência das variáveis de processo na alcoólise enzimática de óleo de mamona. Ciência e Tecnologia de Alimentos, 24, 178-182.

16. Yu, D.H., Tian, L., Wu, H., Wang, S., Wang, Y., Ma, D.X. and Fang, X.X. (2010) Ultrasonic Irradiation with Vibration for Biodiesel Production from Soybean Oil by Novozym 435.

Process Biochemistry, 45, 519-525. http://dx.doi.org/10.1016/j.procbio.2009.11.012

17. Oliveira, D., Luccio, M., Faccio, C., Rosa, C.D., Bender, J.P., Lipke, N., Amroginski, C., Dariva, C. and Oliveira, J.D. (2005) Optimization of Alkaline Transesterification of Soybean Oil and Castor Oil for Biodiesel Production. Applied Biochemistry and Biotechnology, 122, 533-560.

18. Maleki, E., Aroua, M.K. and Sulaiman, N.M.N. (2013) Improved Yield of Solvent Free Enzymatic Methanolysis of Palm and Jatropha Oils Blended with Castor Oil. Applied Energy, 104, 905-909. http://dx.doi.org/10.1016/j.apenergy.2012.12.009

19. Chen, J. and Wu, W.T. (2003) Regeneration of Immobilized Cândida antarctica Lipase for Transesterification. Journal of Bioscience and Bioengineering, 95, 466-469.http://dx.doi.org/10.1016/S1389-1723(03)80046-4

20. Tupufia, S.C., Jeon, Y.J., Marquis, C., Adesina, A.A. and Rogers, P.L. (2013) Enzymatic Conversion of Coconut Oil for Biodiesel Production. Fuel Processing Technology, 106, 721-726. http://dx.doi.org/10.1016/j.fuproc.2012.10.007

21. Abigor, R.D., Uadia, P., Foglia, T.A., Haas, M.J., Jones, K.C., Okpef'a, E., Obibuzor, J.U. and Bafor, M.E. (2000) Lipase-Catalysed Production of Biodiesel Fuel from Some Nigerian Lauric Oils. Biochemical Society Transactions, 28, 979-981. http://dx.doi.org/10.1042/BST0280979

22. Sun, J., Yu, B., Curran, P. and Liu, S.Q. (2012) Lipase-Catalysed Transesterification of Coconut Oil with Fusel Alcoholsin a Solvent-Free System. Food Chemistry, 134, 89-94.http://dx.doi.org/10.1016/j.foodchem.2012.02.070

# Citations

## CHAPTER 1

Nour Sh. El-Gendy, A. Hamdy, and Salem S. Abu Amr, An Investigation of Biodiesel Production from Wastes of Seafood Restaurants, http://dx.doi.org/10.1155/2014/609624.

## CHAPTER 2

Hanifa Taher, Sulaiman Al-Zuhair, Ali H. Al-Marzouqi, Yousef Haik, and Mohammed M. Farid, "A Review of Enzymatic Transesterification of Microalgal Oil-Based Biodiesel Using Supercritical Technology,"Enzyme Research, vol. 2011, Article ID 468292, 25 pages, 2011. doi:10.4061/2011/468292.

# CHAPTER 3

Bernardo Dias Ribeiro, Aline Machado de Castro, Maria Alice Zarur Coelho, and Denise Maria Guimarães Freire, "Production and Use of Lipases in Bioenergy: A Review from the Feedstocks to Biodiesel Production," Enzyme Research, vol. 2011, Article ID 615803, 16 pages, 2011. doi:10.4061/2011/615803.

# CHAPTER 4

Rudras Baliga and Susan E. Powers, Sustainable Algae Biodiesel Production in Cold Climates, doi:10.1155/2010/102179.

# CHAPTER 5

Nina Kolesárová, Miroslav Hutňan, Igor Bodík, and Viera Špalková, Utilization of Biodiesel By-Products for Biogas Production, http://dx.doi.org/10.1155/2011/126798.

# CHAPTER 6

Omotola Babajide, "Sustaining Biodiesel Production via Value-Added Applications of Glycerol,"Journal of Energy, vol. 2013, Article ID 178356, 7 pages, 2013. doi:10.1155/2013/178356.

# CHAPTER 7

Guanyi Chen, Liu Zhao, Yun Qi, and Yuan-Lu Cui, "Chitosan and Its Derivatives Applied in Harvesting Microalgae for Biodiesel Production: An Outlook," Journal of Nanomaterials, vol. 2014, Article ID 217537, 9 pages, 2014. doi:10.1155/2014/217537

# CHAPTER 8

Ribeiro, L., Silva, A., Silva, M., and G. Almeida, R. (2014) A Study on Ethanolysis and Methanolysis of Coconut Oil for Enzymatically Catalyzed Production of Biodiesel. Journal of Sustainable Bioenergy Systems, 4, 215-224. doi:10.4236/jsbs.2014.44020.

# Index